ORGANIC CHEMISTRY 1 LECTURE GUIDE 2017: Color Edition

AF887502

BY RHETT C. SMITH, PH.D.

Marketed by Ideal Quill Publishing Group, LLC
© 2006, 2011-2017

All rights reserved. No part of this book may be reproduced or distributed, in any form or by any means, without permission in writing from the Executive Editor. This includes but is not limited to storage or broadcast for online or distance learning courses.

Printed in the United States of America

10 9 8 7 6 5 4 3 2 1

ISBN 978-0-692-89758-4

Organic Chemistry 1 Lecture Guide 2017:

Color Edition

By Rhett C. Smith, Ph.D.

© 2006, 2011-2017

Contents

PART II: Substitution, Elimination and Oxidation

Lecture Topics in Part II:

1. Stereochemistry V: Tracking Stereocenters in Reactions
2. Nucleophilicity versus Basicity
3. Substitution Reactions of RX I: The S_N1 Mechanism
4. Substitution Reactions of RX II: The S_N2 Mechanism
5. Substitution Reactions of RX III: Stereochemistry of Substitution Reactions
6. Elimination Reactions of RX I: The E1 Mechanism
7. Elimination Reactions of RX II: The E2 Mechanism
8. Stereoelectronic Effects in the E2 Reaction
9. Factors Leading to Non-Zaitsev Products in the E2 Reaction: Hofmann Elimination
10. Competition Among Substitution and Elimination Reactions of RX: Predicting Pathways
11. Reactions of Alcohols I: Substitution and Elimination
12. Oxidation and Reduction: Definitions
13. Reactions of Alcohols II: Oxidation
14. Acid Cleavage of Ethers
15. Ring-Opening of Epoxides: Steric versus Electronic Effects on Product Distribution

Part III: Reactions Involving Alkenes, Alkynes and Radicals

Lecture Topics in Part III:

1. Nomenclature II: Alkenes and the E-/Z- Convention
2. Properties of Alkenes and Hydrogenation of Alkenes
3. Electrophilic Addition to Alkenes and Carbocation Rearrangement Revisited
4. Markovnikov's Rule, Syn- and Anti- Addition
5. Reactions of Alkenes I: Hydrohalogenation and Hydration
6. Reactions of Alkenes II: Halogenation and Halohydrin Formation
7. Reactions of Alkenes III: Oxymercuration/Reduction
8. Reactions of Alkenes IV: Cyclopropanation and Epoxidation

PART I: Structure, Stability and Conventions

Lecture Topics in Part I:

1. Coulombic Forces and Representation of Structure
2. Polarity, Dipole Moments and Formal Charge
3. Intermolecular Forces, Boiling Point and Melting Point
4. Nucleophiles, Electrophiles and Functional Groups
5. Arrow-Pushing Formalism
6. Hybridization, Sigma and Pi Bonds, Lone Pairs and Bond Geometry
7. Resonance and Delocalization Energy
8. Elementary Steps of Reaction Mechanisms: Applying the Arrow-Pushing Formalism
9. Acids and Bases I: Definitions
10. Acids and Bases II: Relating Structure to the Strength of an Acid or Base
11. Stability of Cations and Radicals
12. Predicting Reaction Spontaneity and Direction of Equilibria
13. Reaction Coordinate Diagrams and Reaction Rate
14. Nomenclature of Alkanes
15. Cycloalkanes and Configurational Isomerism
16. Nomenclature of Alkyl Halides and Alcohols
17. Isomerism and Conformational Analysis I: Linear Alkanes and Newman Projections
18. Conformational Analysis II: Cycloalkanes and the Chair Conformation of Cyclohexane
19. Stereochemistry I: Chirality and Optical Activity
20. Stereochemistry II: Cahn-Ingold-Prelog Rules and Assigning R- and S- Labels of Configuration
21. Stereochemistry III: Fischer Projections

Lecture Topic I.1: Coulombic Forces and Representation of Structure
Coulombic Forces

1. Opposite Charges Attract

2. Like Charges Repel

__Notes__

Lecture Topic I.1: Coulombic Forces and Representation of Structure
Review of Lewis Dot Structures

Drawing Lewis Dot Structures:

When given the formula for a molecule or ion:

1. Add up valence electrons; if it is an ion and has a charge, add an electron for each "–" charge, take away an electron for each "+" charge.
2. Draw single bonds between the atoms
3. Add remaining valence electrons to the atoms to give them the expected number of lone pairs (you know this from the Lewis structure of the given element)
4. Add bonds between adjacent elements that have open valences, making multiple bonds.

HINT:

Most neutral species we will encounter in this class feature elements with a total of 8 electrons (an **octet**) around them in the correct structure, except H, which will have only 2.

Notes

Lecture Topic I.1: Coulombic Forces and Representation of Structure
Lewis Dot Structure Examples

	Target	Atoms	$\#e^-$	Structure
A.	CCl_4	C 4×Cl		
B.	CH_3OH	C 4×H O		
C.	CH_2O	C 4×H O		

Notes

Lecture Topic I.1: Coulombic Forces and Representation of Structure

Timesavers: Shorthand Representations of Molecules

dash formula	condensed formula	bond-line formula
(A) 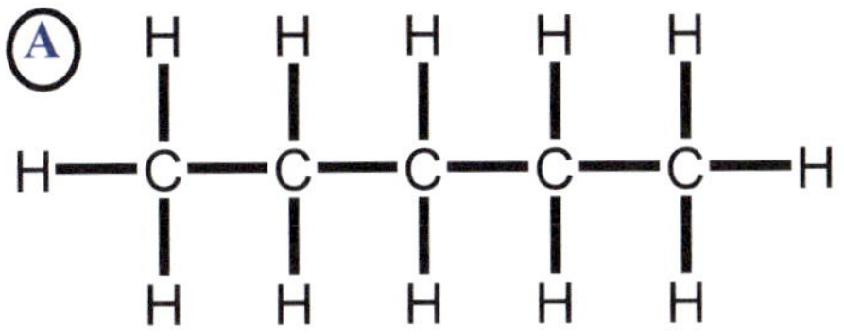	$CH_3CH_2CH_2CH_2CH_3$ $CH_3(CH_2)_3CH_3$	
(B) 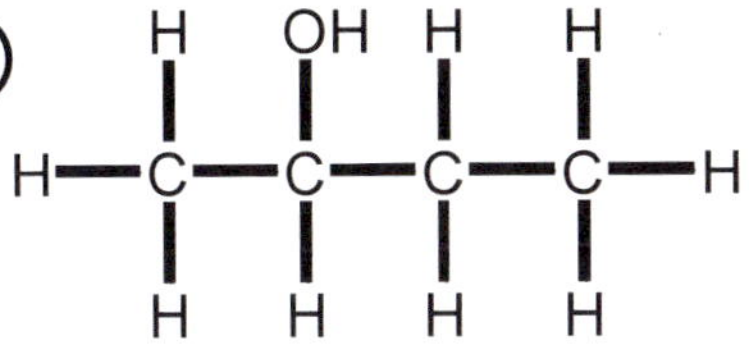	$CH_3CH(O)CH_2CH_3$	
(C) 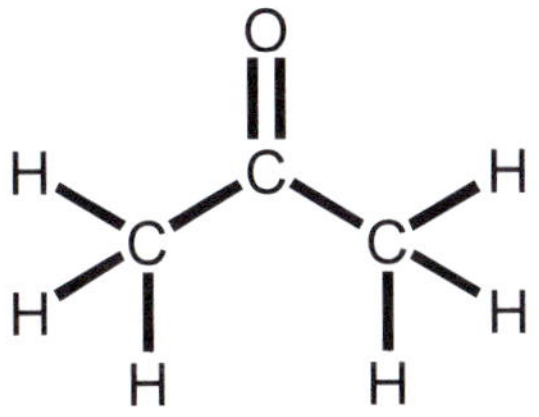	$CH_3C(O)CH_3$	

Notes

Lecture Topic I.1: Coulombic Forces and Representation of Structure
Representing the 3D Shape of Molecules

Since orbitals point in different directions, it should not be surprising that bonds point in various directions, and that molecules are three-dimensional. Chemists use a simple shorthand, called the "dash-wedge notation" in order to attempt to indicate a molecules 3D shape on 2D paper. It's not too complicated ...

A "regular" line indicates a bond that lies

(A)

A "wedge" indicates a bond that is

(B)

A "dashed wedge" indicates a bond that is

(C)

Notes

Lecture Topic I.2: Polarity, Dipole Moments and Formal Charge
Defining Electronegativity

ionic bonds: "Complete" transfer of electrons to give two ions

(A)

cation **anion**

covalent bonds: Equal or Unequal sharing of electrons without "complete" transfer

(B)

A bond can be represented by two dots or by a line.

A line represents two electrons!

The extent to which the electron is transferred depends upon how strongly each atom attracts the electrons within the bond. This attractive force is called the ELECTRONEGATIVITY of the atoms.

Electronegativity (EN):______________________________.

increases________________________in the periodic table

Notes

Lecture Topic I.2: Polarity, Dipole Moments and Formal Charge

Polar and Ionic Bonds

The difference in electronegativity (EN) of two bonded atoms will tell you to what extent the electrons in the bond are transferred from one atom to the other.

H—H, $EN = 2.1 \;\; 2.1$
Cl—Cl, $EN = 3.0 \;\; 3.0$
} $\Delta EN = 0$ equal sharing of electrons
= ______________________.

$\delta^+ \;\; \delta^-$
H—Cl, $EN = 2.1 \;\; 3.0$ $\Delta EN = 0.9$ unequal sharing of electrons
= ______________________.

Na^+Cl^-, $EN = 0.9 \;\; 3.0$ $\Delta EN = 2.1$ transfer of electrons
= ______________________.

general rule: ΔEN < ____, = "covalent"
> ____, = "ionic"

Δ = uppercase "delta". In science and math, delta often represents a change or difference.

Notes

Lecture Topic I.2: Polarity, Dipole Moments and Formal Charge

Representing Polar Bonds

δ^+, δ^-

These are lowercase "delta" symbols. In science and math, delta often represents a change or a difference.

(A) ______________________________ these lowercase Greek symbols ('delta') tell us that the atom above which they are written bear a partial positive or negative charge.

$\overset{\delta^+}{H}$—$\overset{\delta^-}{F}$

The dipole moment is represented by 'mu', the greek 'm'.

(B) An arrow can also be used to represent the pull of electrons in the direction of the arrow. To emphasize the direction of electron movement, a 'plus' sign is drawn at the other side of the arrow. These arrows show us the

____

of the molecule.

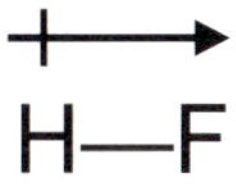

Notes

Formal Charge: Review and Examples

Formal charge = ________________________________

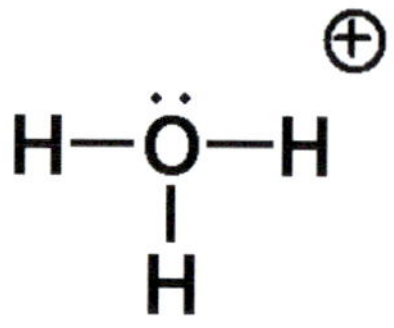

O:

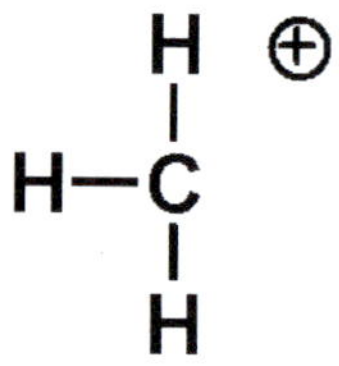

C:

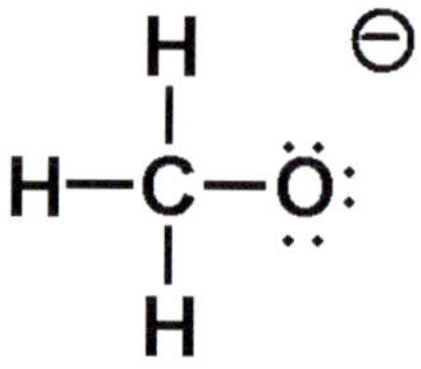

C:
O:

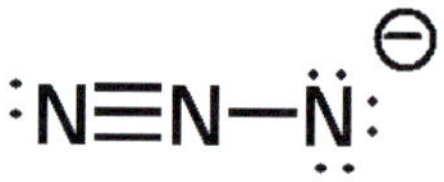

1 2 3

N#1:
N#2:
N#3:

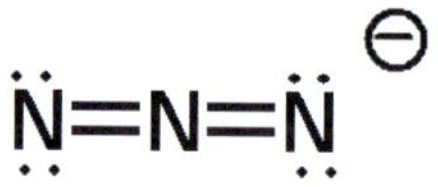

1 2 3

N#1:
N#2:
N#3:

C:
O:

Notes

Lecture Topic I.3: Intermolecular Forces, Boiling Point and Melting Point

Intermolecular Forces

There are three types of intermolecular forces of focus in organic chemistry:

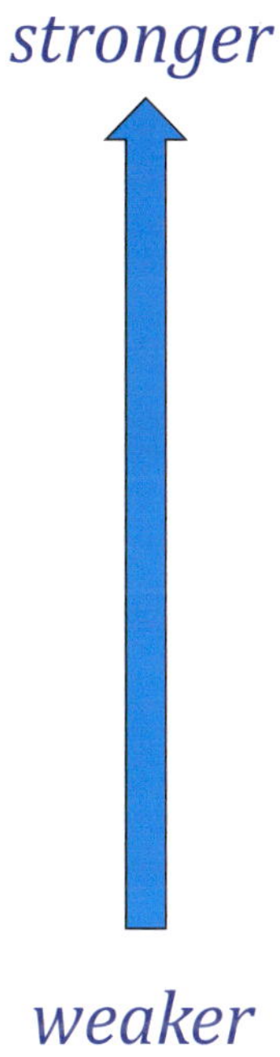

i. *Hydrogen Bonding*

ii. *Dipole-Dipole interactions*

iii. *van der Waals interactions* (London dispersion forces).

Notes

Lecture Topic I.3: Intermolecular Forces, Boiling Point and Melting Point
Predicting Melting and Boiling Points

In a solid or liquid, it is the intermolecular forces that hold molecules close together:

In a molecular solid, energy is needed to break the molecules apart so they may flow past one another to form a liquid (melting).

In a molecular liquid, energy is needed to give the molecules enough kinetic energy to allow them to break away from the liquid surface and enter the gas phase (boiling).

It takes more energy to disrupt stronger intermolecular forces, so:

Melting point and boiling point both ______________________as the

strength of intermolecular forces_______________________________.

Notes

Lecture Topic I.4: Nucleophiles, Electrophiles and Functional Groups
Nucleophiles and Electrophiles

Nucleophile:

example:

Electrophile:

example:

Notes

Lecture Topic I.4: Nucleophiles, Electrophiles and Functional Groups
Functional Groups I: Hydrocarbons

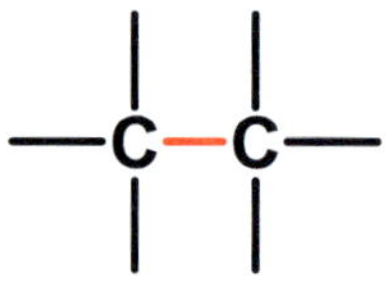

Alkane

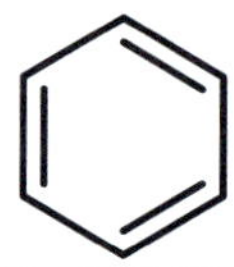

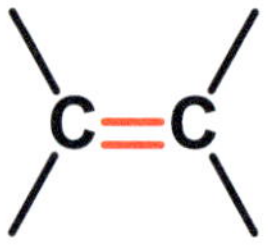

Alkene

Alkyne

17

Notes

Lecture Topic I.4: Nucleophiles, Electrophiles and Functional Groups

Functional Groups II: Containing Oxygen, Nitrogen and Halogens

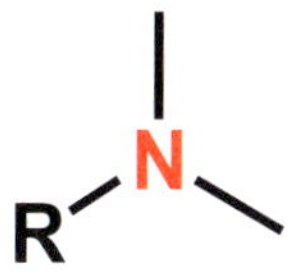

R—OH

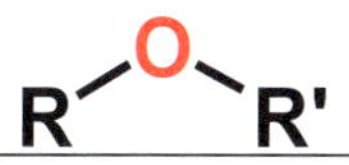

R—X

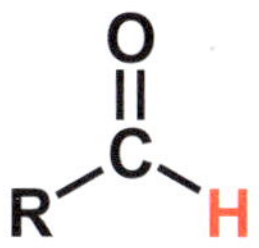

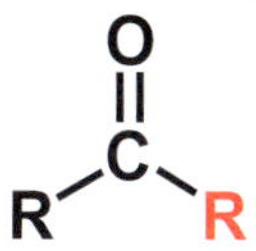

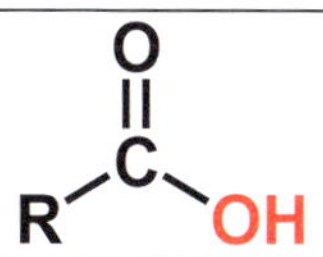

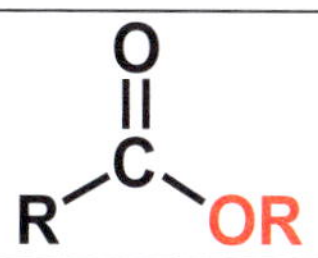

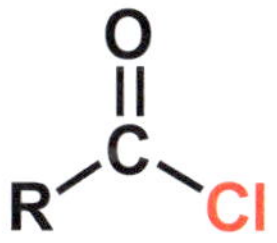

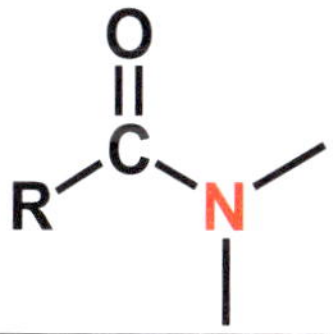

Notes

Lecture Topic I.4: Nucleophiles, Electrophiles and Functional Groups
Identifying Functional Groups in Organic Molecules

Circle and label each non-alkane functional group in the structure of Taxol, a chemotherapeutic drug:

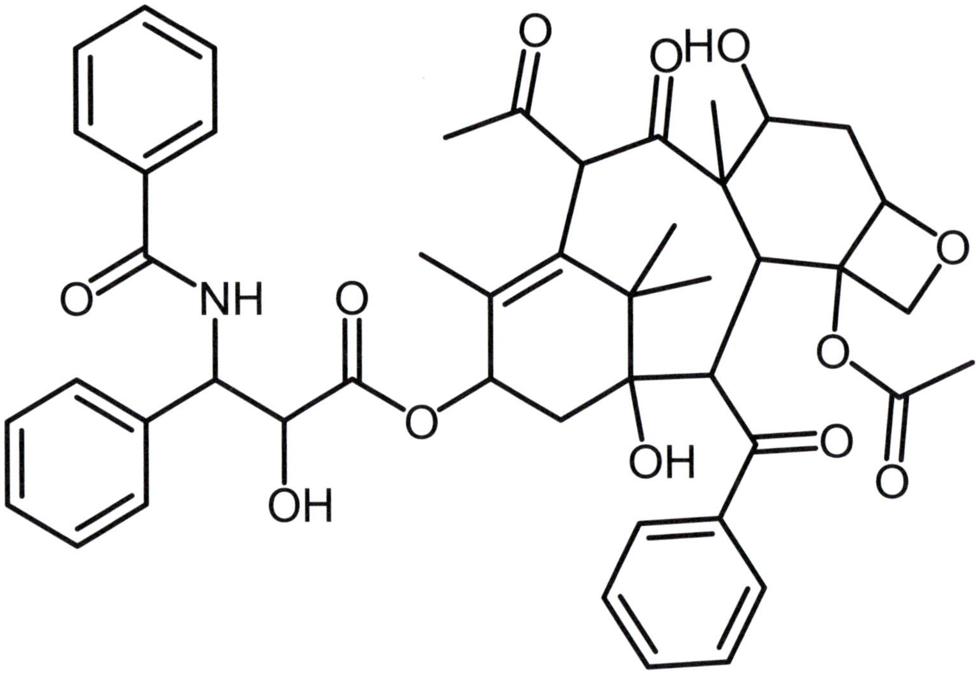

Notes

Lecture Topic I.5: Arrow-Pushing Formalism

Representing Transfer of Electrons

Since bonds consist of electrons between nuclei, the movement of electrons leads to changes in bonding; in other words **Chemical Reactions** are described by showing how the electrons rearrange upon going from reactants to products. In order to show the movement of electrons, we need a notation that is clear.

(A)

(B)

Electrons move FROM a good electron source (a lone pair or an anion, or the negative end of a polar bond ...) TO a good electron acceptor or "sink" (the positive end of a polar bond, an empty orbital, a cation ...).

Notes

Lecture Topic I.5: Arrow-Pushing Formalism

Some Arrow-Pushing Examples

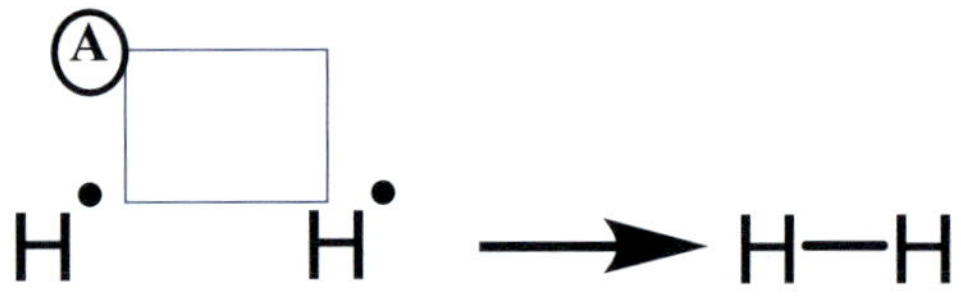

Two hydrogen atoms react to form H_2; each gives one electron, so use fishhook arrows

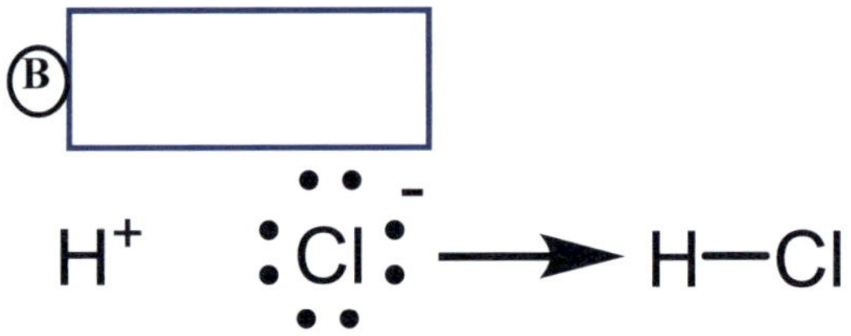

A chloride anion and a proton react to form HCl; chloride donates a pair of electrons, so use a standard arrow.

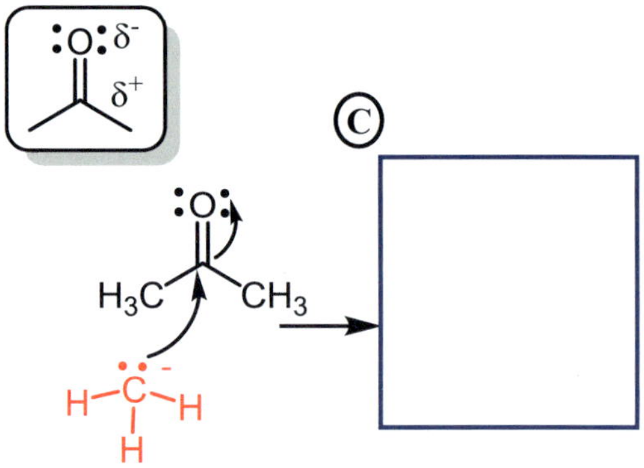

Here we show a more complicated 'cascade' of two arrows. The methyl anion electron pair (a good electron source) attacks the positive end of the polar C=O bond. That carbon already has 8 electrons, so 2 electrons have to move away from carbon, towards the more electronegative O atom that has a pull for electrons.

Notes

Lecture Topic I.6: Hybridization, Sigma and Pi Bonds, Lone Pairs and Bond Geometry

Atomic Orbitals

l	0	1	2
Name	**s**	**p**	**d**
Shapes		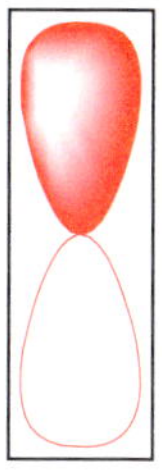	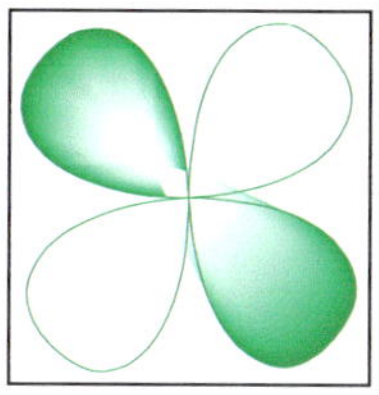

s orbitals are spherical

p orbitals are hourglass-shaped
There are three *p* orbitals in each shell;
One on each of the axes (x, y, and z)

Notes

Lecture Topic I.6: Hybridization, Sigma and Pi Bonds, Lone Pairs and Bond Geometry

Hybrid Orbitals

Two Orbitals can combine ...

... by adding ...

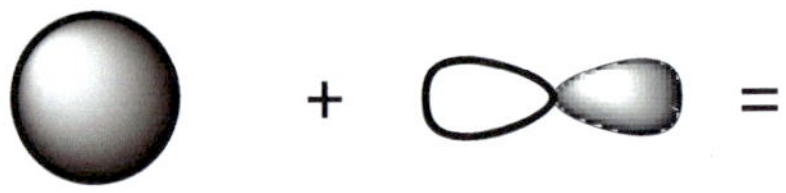

s orbital   *p* orbital

(A)

... or by subtracting ...

s orbital   *p* orbital

(B)

... to form **two** new orbitals.

Adding or "hybridizing" one *s* orbital with one *p* orbital gives two ______ hybrid orbitals.

Notes

Lecture Topic I.6: Hybridization, Sigma and Pi Bonds, Lone Pairs and Bond Geometry

The sp-Hybridized Atom

Drawing both sp hybrid orbitals on the same atom:

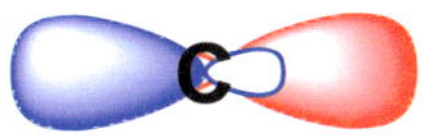

This is an ______________________________ carbon atom.

The *sp* orbitals are pointed directionally ___________

from one another, resulting in a _____________ geometry.

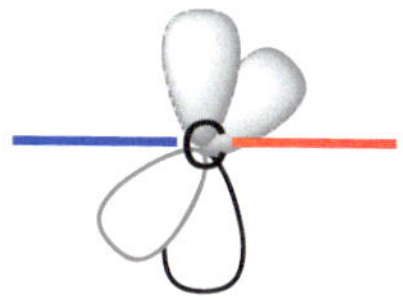

Since carbon has three *p* orbitals and we only used one to make the hybrids, two *p* orbital are still left unhybridized on carbon:

__Notes__

Lecture Topic I.6: Hybridization, Sigma and Pi Bonds, Lone Pairs and Bond Geometry

The sp^2-Hybridized Atom

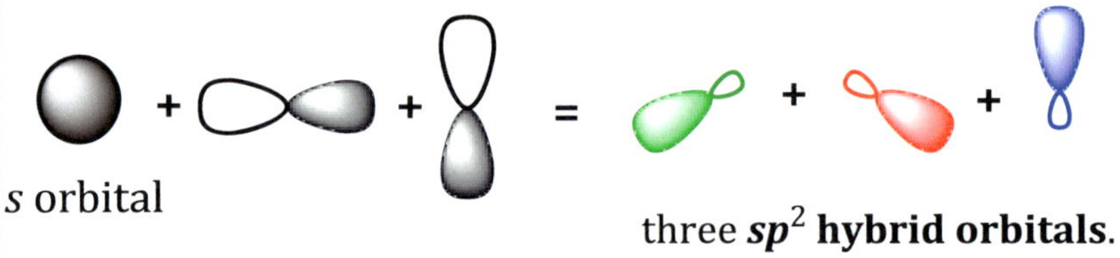

Drawing all three sp^2 hybrid orbitals on the same atom:

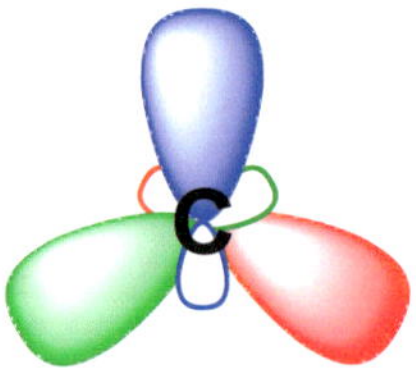

This is an **sp^2 hybridized** carbon atom. The sp^2 orbitals are in a plane pointed directionally

____

from one another, towards the corners of an equilateral triangle in a

____ geometry.

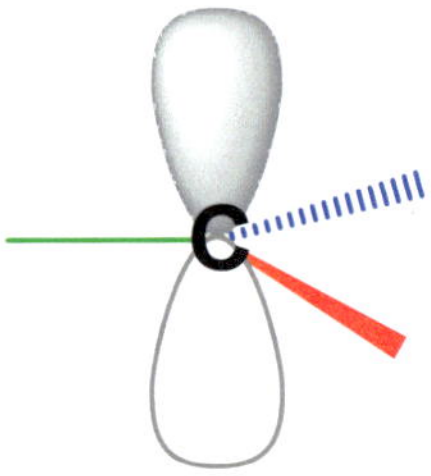

Since carbon has three *p* orbitals and we only used two to make the hybrids, a p orbital is still left unhybridized on carbon.

Notes

Lecture Topic I.6: Hybridization, Sigma and Pi Bonds, Lone Pairs and Bond Geometry

The sp³-Hybridized Atom

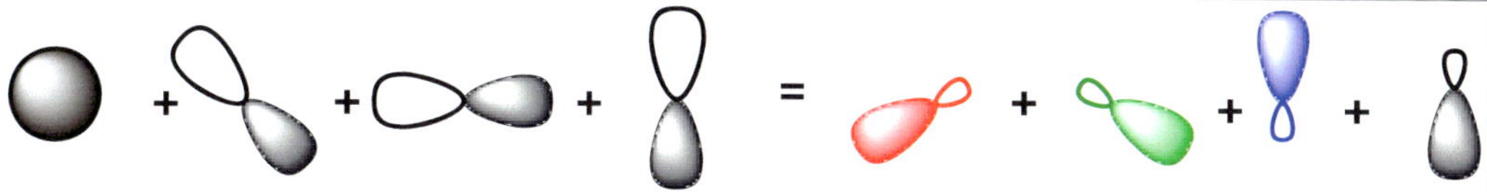

s orbital — three *p* orbitals — Four ***sp*³ hybrid orbitals**.

Drawing all four sp³ hybrid orbitals on the same atom:

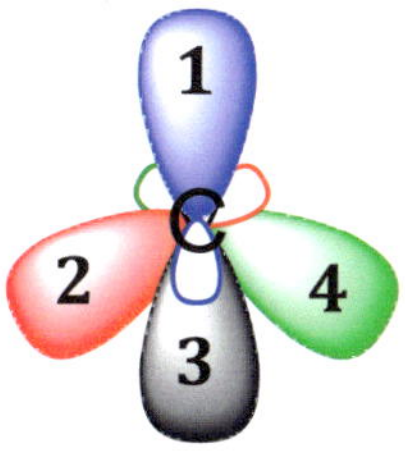

This is an *sp*³ hybridized carbon atom. The sp³ orbitals are pointed directionally ______________________________ from one another, in a ____ geometry.

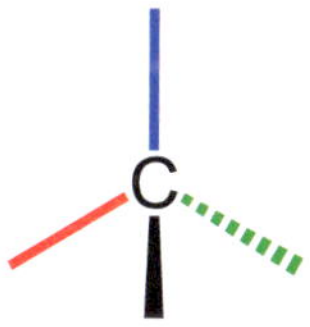

Since carbon has three p orbitals and we used all three to make the hybrids, all of the valence orbitals are now *sp*³ orbitals, and consequently these are the only ones used in bonding.

Notes

Lecture Topic I.6: Hybridization, Sigma and Pi Bonds, Lone Pairs and Bond Geometry

Sigma (σ) and Pi (π) Bonds

A sigma (σ) bond is one in which the electrons making the bond are between the two nuclei joined by the bond:

A pi (π) bond is one in which the electrons lie above and below the line between the two nuclei which are joined by the bond:

In the vast majority of examples in introductory organic, the **hybrid orbitals** will form sigma bonds or will be filled by a lone pair of electrons, and any pi bonds will be made by p orbitals not hybridized into sp, sp^2, or sp^3 orbitals…

Notes

Lecture Topic I.6: Hybridization, Sigma and Pi Bonds, Lone Pairs and Bond Geometry

Hybridization and Potential for Bonding

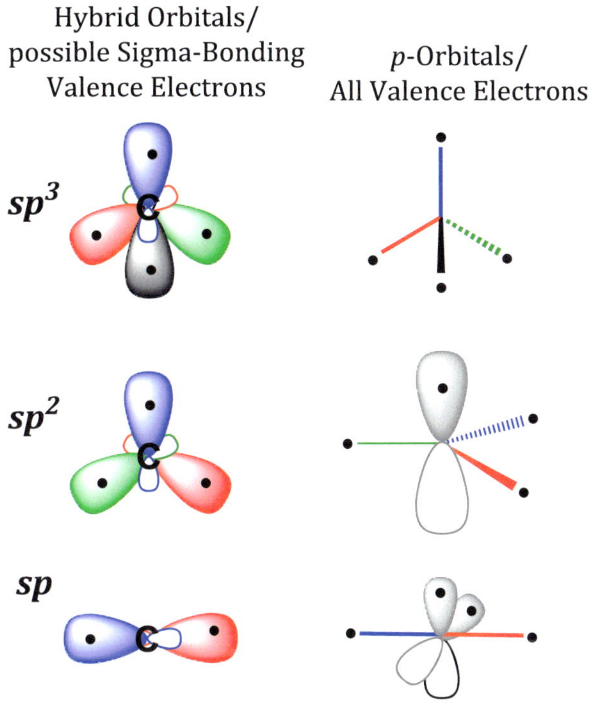

Notes

Lecture Topic I.6: Hybridization, Sigma and Pi Bonds, Lone Pairs and Bond Geometry

Bonding in Ethane (C_2H_6)

We can bind two sp^3 hybridized carbons together and make sigma bonds to hydrogen atoms with the remaining sp^3 orbitals

This is ETHANE. There are no more valence orbitals, so we can make no more bonds.

__Notes__

Lecture Topic I.6: Hybridization, Sigma and Pi Bonds, Lone Pairs and Bond Geometry

Bonding in Ethylene (C_2H_4)

In Ethylene, the *sp*^{2} hybrid orbitals form sigma bonds with C and H.

We have a pi orbital left over on each carbon, each with one electron. These two orbitals overlap to form a pi bond, so ethylene's line-bond structure is:

Notes

Lecture Topic I.6: Hybridization, Sigma and Pi Bonds, Lone Pairs and Bond Geometry

Bonding in Acetylene (C_2H_2)

In acetylene, the *sp*-hybrid orbitals are again used to make the sigma bonds.

Each of the two *sp*-hybridized carbon atoms has two *p* orbitals left unhybridized, so there will be two pi bonds formed between the carbon atoms, yielding acetylene's line-bond structure:

__Notes__

Lecture Topic I.6: Hybridization, Sigma and Pi Bonds, Lone Pairs and Bond Geometry

The Influence of Hybridization on Electronegativity

Electronegativity is how strongly an atom holds its electrons in a bond. How strongly electrons are held depends on the orbital that holds the electrons:

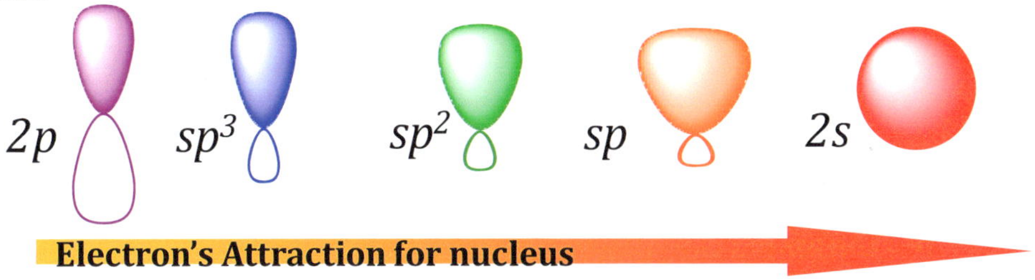

An orbital that has more (A) [ ] has a greater pull on electrons.

So, the electronegativity trend is:

(B) [ ]

An *sp* hybridized C atom has about the **same electronegativity** as an *sp*^{3}-hybridized N atom

Notes

Lecture Topic I.7: Resonance and Delocalization Energy

Review of Resonance

In a resonance structure the atomic positions remain the same, but electrons move around (lone pairs and multiple bond positions change).

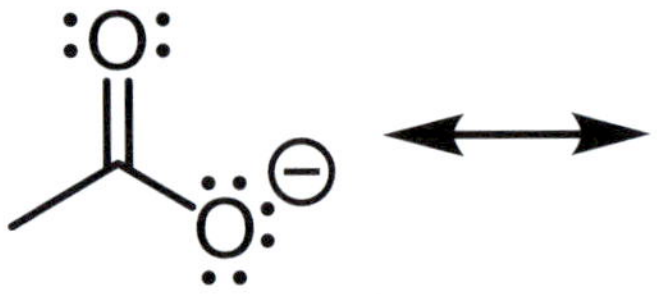

Ⓐ

Ⓑ

neither resonance form alone accurately describes the anion; the "real" structure is a combination of the two, as shown at right

Resonance Hybrid
(dashed line = partial bond)

Why do we care? Resonance structures can help us understand reactivity. If a molecule can be drawn in numerous resonance forms, it has some added stability known as:

This extra stability is due to:

Notes

Lecture Topic I.7: Resonance and Delocalization Energy
Representing Resonance Delocalization

In the previous example, two plausible Lewis Structures could be drawn for acetate. We can use arrow-pushing to indicate how two **resonance contributors** are related. You should also be able to fill in the formal charge on each atom and curved arrows to show electron movement:

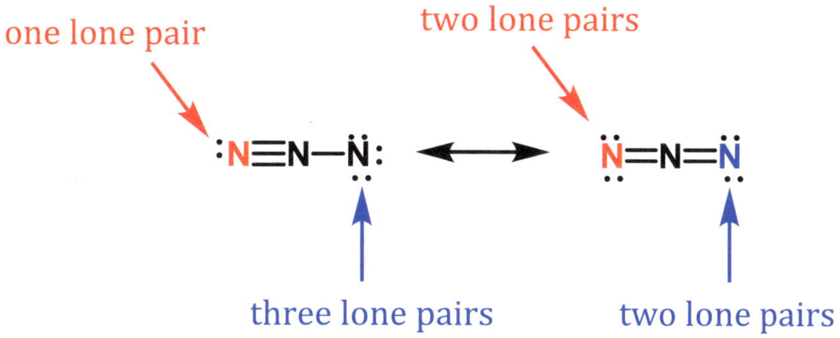

Note the special resonance arrow

Not to be confused with the equilibrium arrow!
RESONANCE IN NOT AN EQUILIBRIUM!

Notes

Lecture Topic I.8: Elementary Steps of Reaction Mechanisms
Common ways electrons move

We are not yet attempting to predict when or explain why these steps take place, but you should be able to 1. recognize which step is happening given reactants/reagents; 2. provide the arrows necessary for a transformation or 3. provide product(s) given reagent(s) and arrow(s).

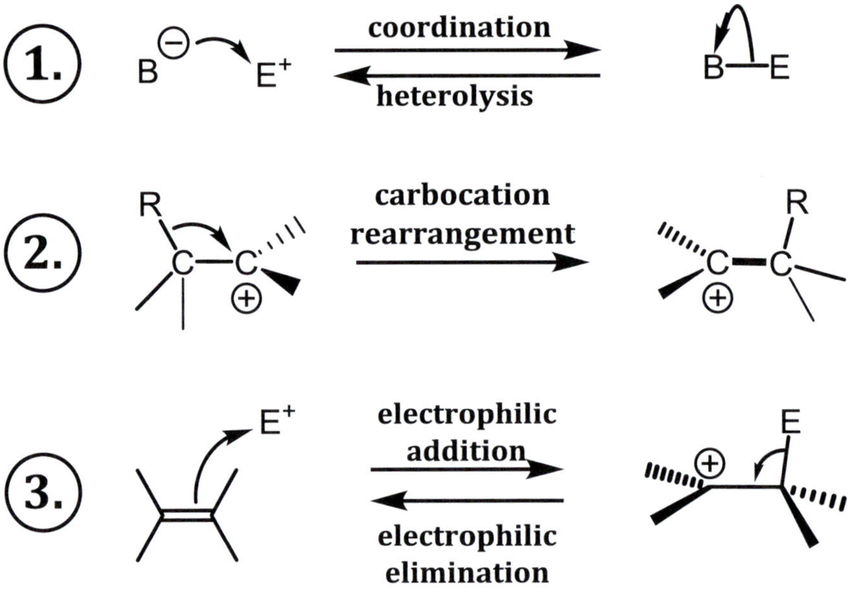

Notes

Lecture Topic I.8: Elementary Steps of Reaction Mechanisms
Common ways electrons move

4.

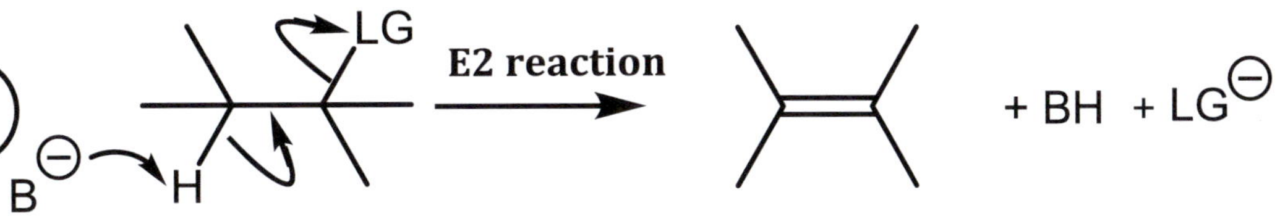

5.

6.

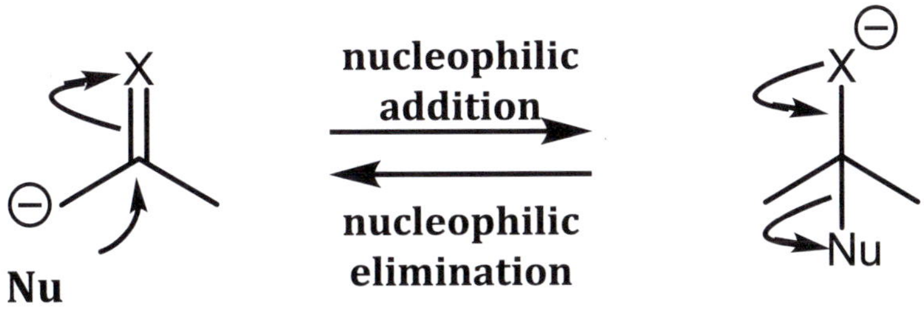

Notes

Lecture Topic I.8: Elementary Steps of Reaction Mechanisms
Coordination/Heterolysis (Including Protonation/Deprotonation)

Here are some examples of coordination and heterolysis. Protonation is a common example of coordination and deprotonation is a common example of heterolysis.

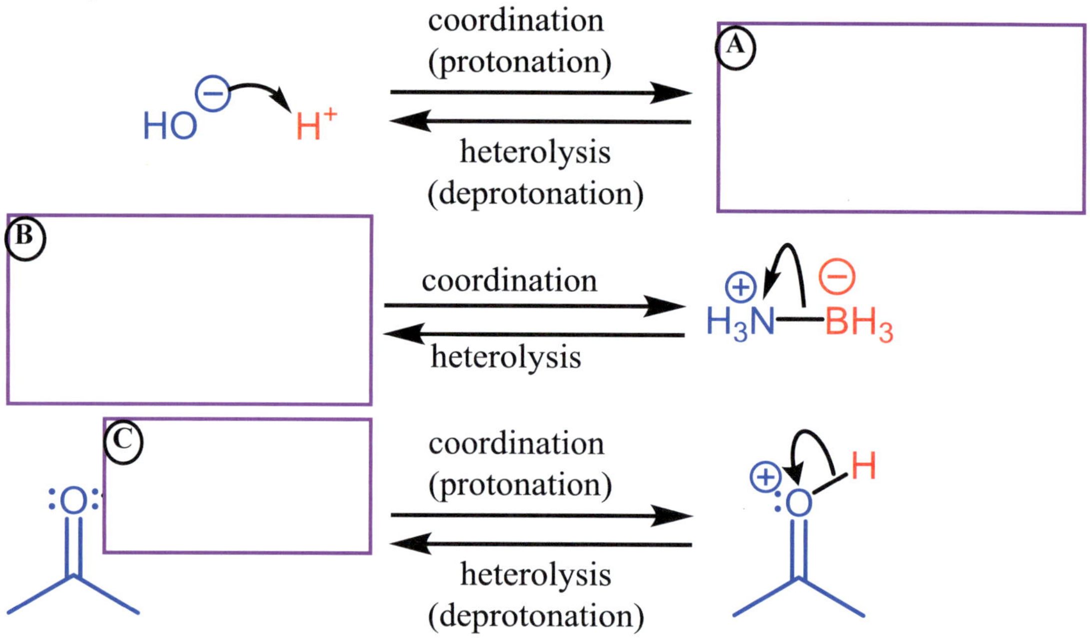

Notes

Lecture Topic I.8: Elementary Steps of Reaction Mechanisms

Carbocation Rearrangement

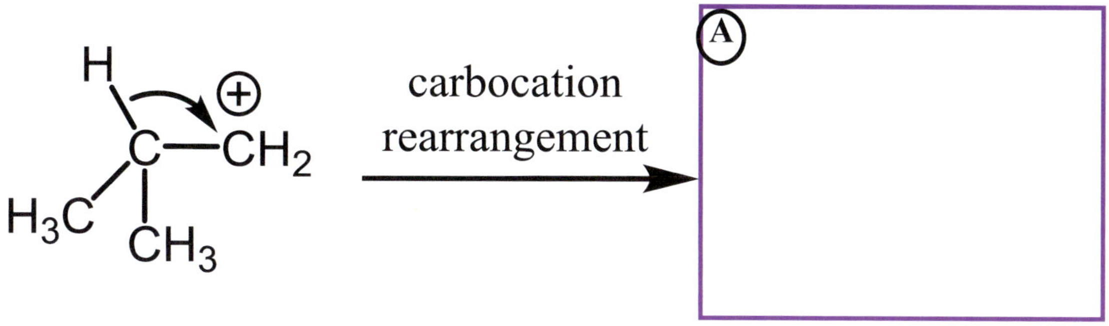

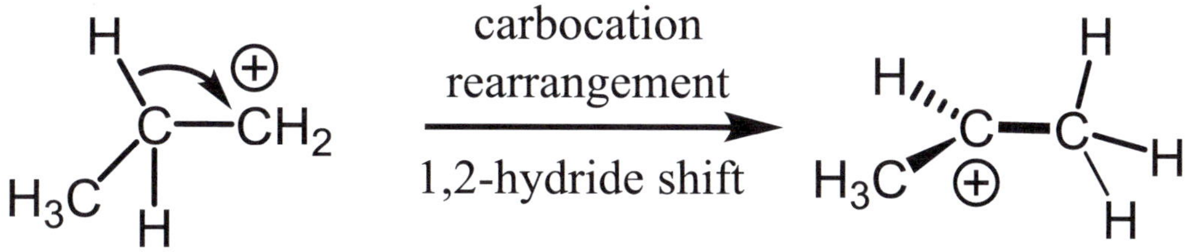

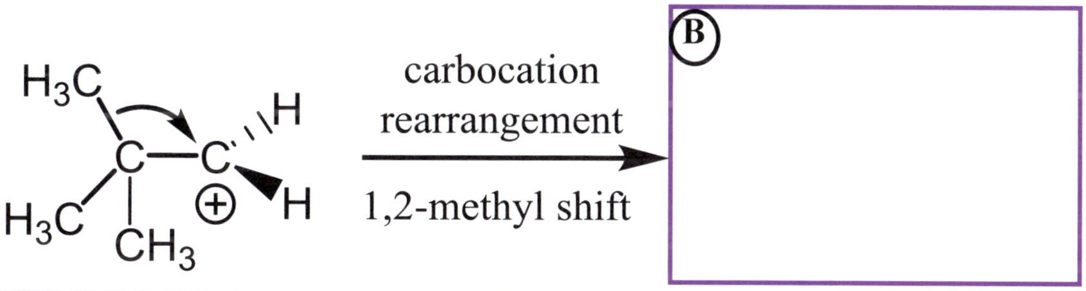

Notes

Lecture Topic I.8: Elementary Steps of Reaction Mechanisms
Electrophilic Addition and Elimination

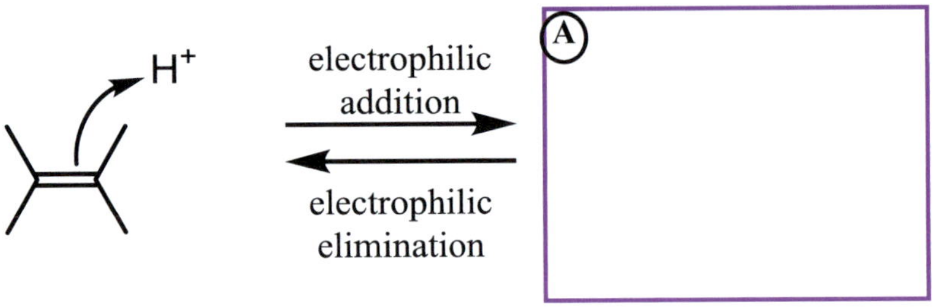

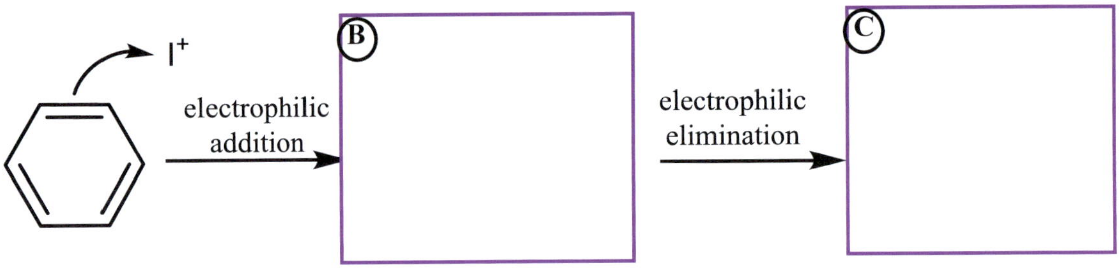

Notes

Lecture Topic I.8: Elementary Steps of Reaction Mechanisms
The E2 Mechanism (Bimolecular Elimination)

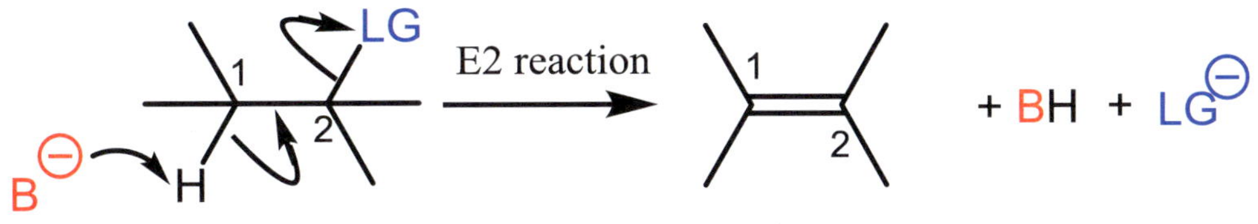

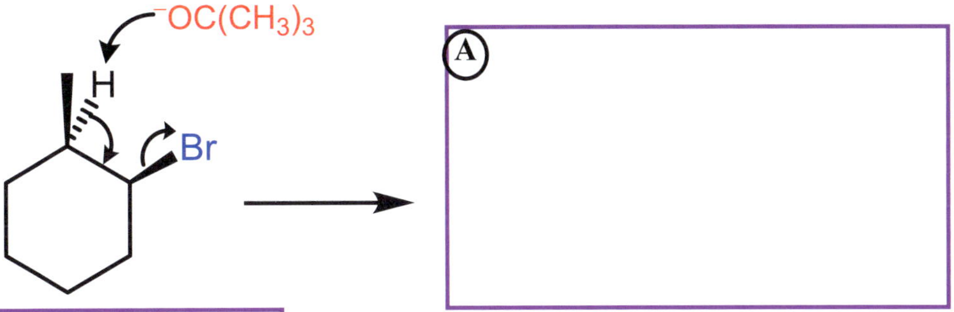

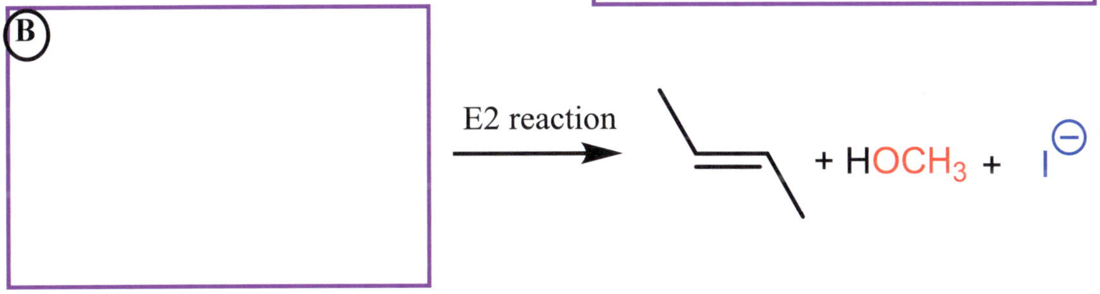

Notes

Lecture Topic I.8: Elementary Steps of Reaction Mechanisms

The S_N2 Mechanism (Bimolecular Substitution)

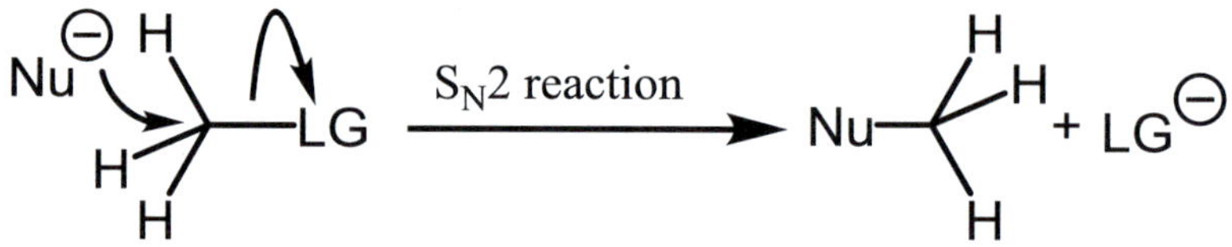

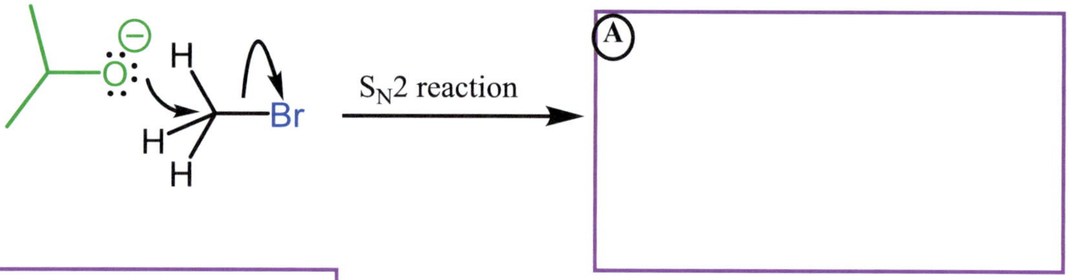

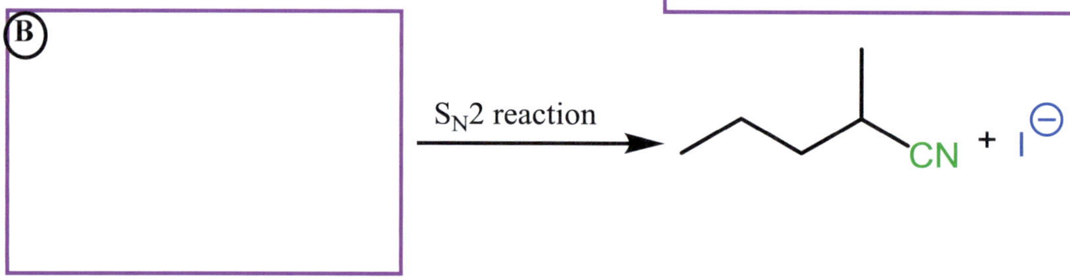

Notes

Lecture Topic I.8: Elementary Steps of Reaction Mechanisms

Nucleophilic Addition and Elimination

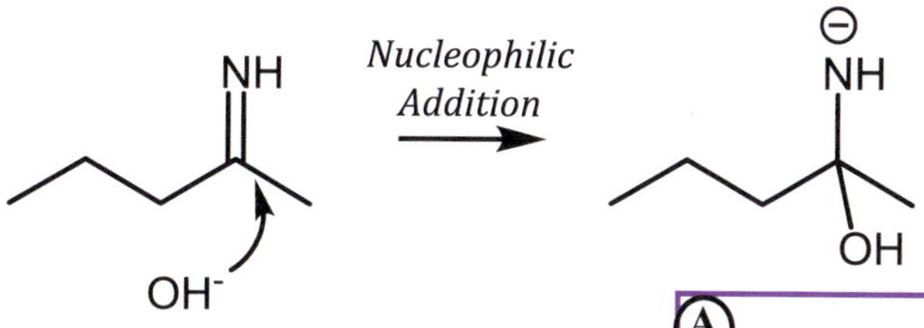

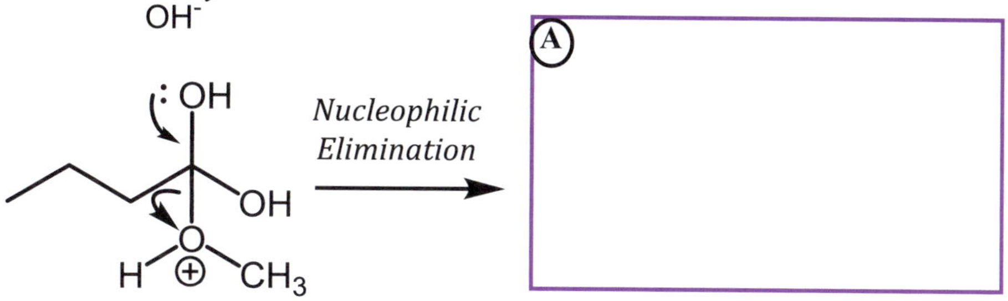

Notes

Lecture Topic I.9: Acids and Bases I: Definitions
Arrhenius, Bronsted-Lowry and Lewis Conventions

In General Chemistry we learned about acids and bases, pH, pK_a, etc. We will briefly review some of these ideas. There are various ways to think about acid-base reactions.

Examples

Arrhenius Definition
acid:
base:

acid:
base:

Brønsted-Lowry Definition
acid:
base:

acid:
base:
base:

Lewis Definition
acid:
base:

acid:
base:

*recall that in water H^+ forms the hydronium ion, H_3O^+

Notes

Lecture Topic I.10: Acids and Bases II

Relating Structure to the Strength of an Acid or Base

$$HA + H_2O \rightleftharpoons A^- + H_3O^+$$

acid — *conjugate base*

Any factors that favor dissociation of HA into H^+ (to form H_3O^+ in water) and A^- will enhance acidity.

Most important factor influencing dissociation:

***1. ______________________________**

Notes

Lecture Topic I.10: Acids and Bases II

Influence of Anionic Atom Electronegativity

The electronegativity plays a strong role because:

CH_4	NH_3	H_2O	HF
________			________
pK_a ~ ______	______	______	______

Notes

Lecture Topic I.10: Acids and Bases II

Influence of Anionic Atom Hybridization on Its Electronegativity

If the atom to be deprotonated is the same and only hybridization changes, we need to know that the hybridization influences electronegativity:

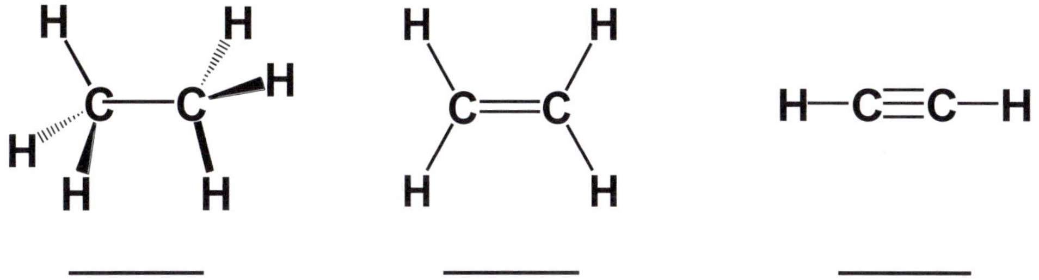

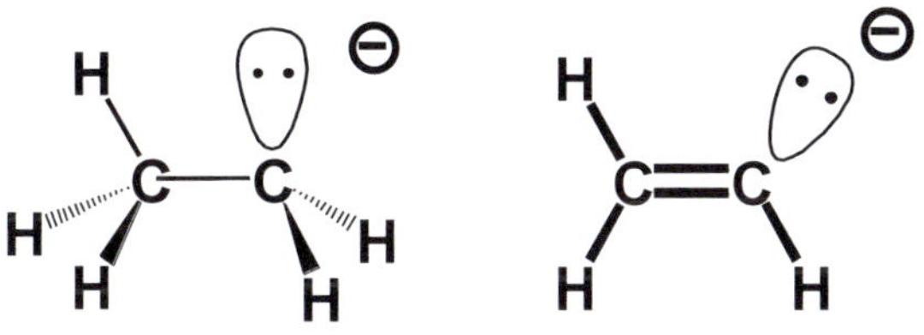

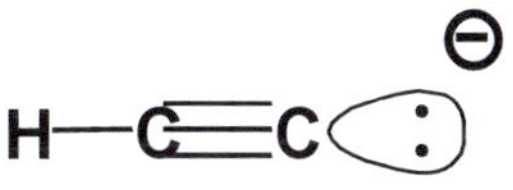

Acetylide anion

Note: the electronegativity of an *sp*-hybridized C is similar to that of an sp^3-hybridized N

Notes

Lecture Topic I.10: Acids and Bases II

Influence of Resonance

If the anion produced by deprotonation has more than one (good) resonance form, then:

HF

pK_a = 3.1

CH_3CO_2H

pK_a = 4.7

CH_3OH

pK_a = 15.5

Notes

Lecture Topic I.10: Acids and Bases II
Influence of Anionic Atom Size

Within a group (column), the size of the anion has a strong effect on acidity, because:

$pK_a \sim$

_______________ **HF** ________

HCl ________

HBr ________

_______________ **HI** ________

- ORBITAL OVERLAP
- **ANION STABILITY**

Notes

Lecture Topic I.10: Acids and Bases II
Inductive Effects that Stabilize Anions

Inductive effects can make a species more OR less acidic. If an atom to be deprotonated has a partial positive charge INDUCED on it by nearby atoms, it is easier to deprotonate because:

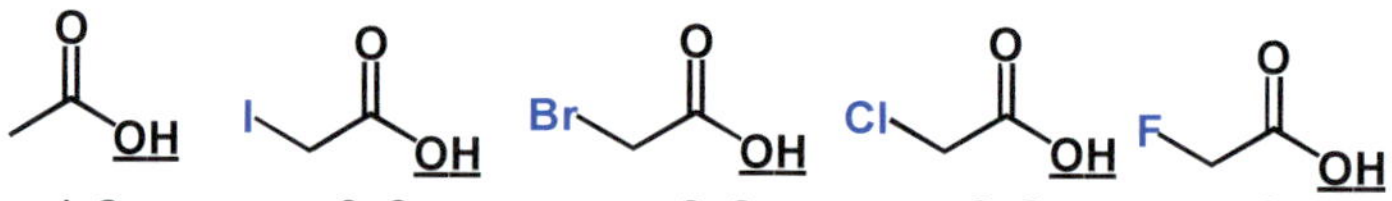

pK_a 4.8 3.2 2.9 2.8 2.7

This series illustrates:

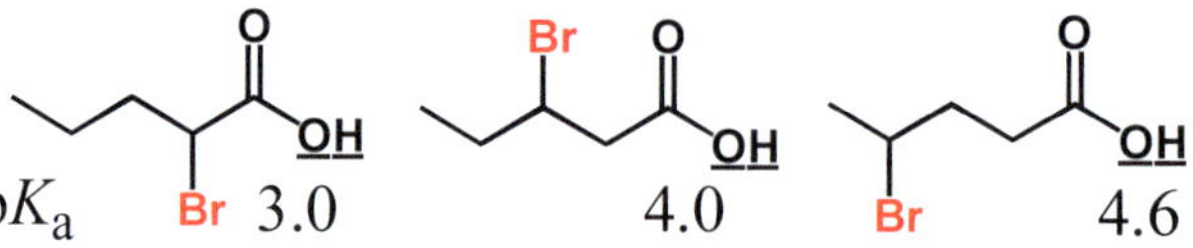

pK_a 3.0 4.0 4.6

This series illustrates:

Notes

Lecture Topic I.10: Acids and Bases II
Inductive Effects that Destabilize Anions

sterics:

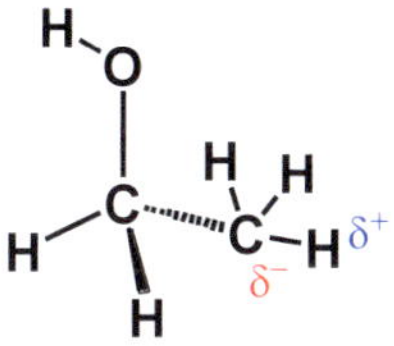 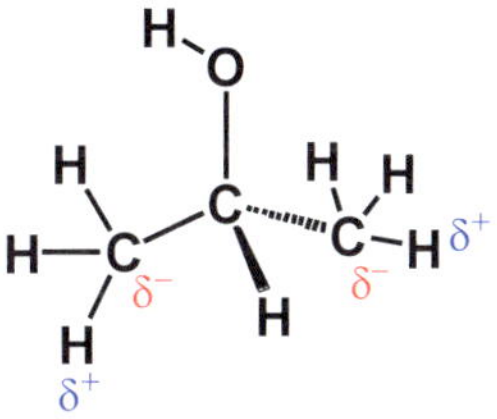 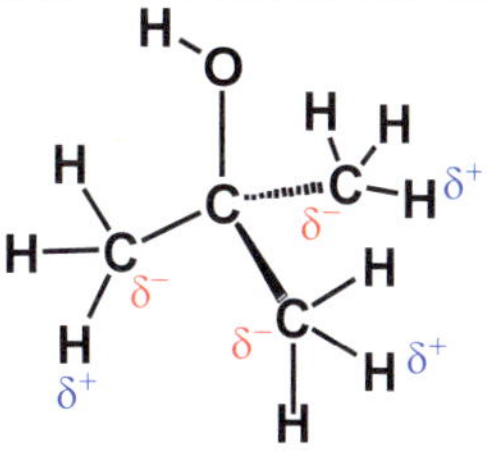

pK_a ~ ______ ______ ______

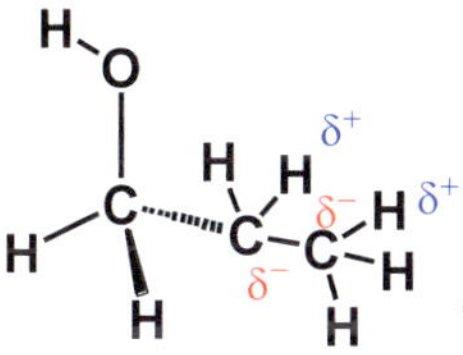 ______

Notes

Lecture Topic I.10: Acids and Bases II
Using Predicted Conjugate Base Anion Stability to Solve Problems

Which is the strongest acid:

a) SiH_4 b) PH_3 c) SH_2 d) HCl

Which is the strongest base:

a) H_3C^- b) H_2N^- c) HO^- d) F^-

Which has the lowest pK_a:

CH_4	$H_3C—CH_3$	$H_2C=CH_2$	$HC≡CH$
(a)	(b)	(c)	(d)

Notes

Lecture Topic I.10: Acids and Bases II

Using Predicted Conjugate Base Anion Stability to Solve Problems

Which proton shown in each molecule is most acidic?

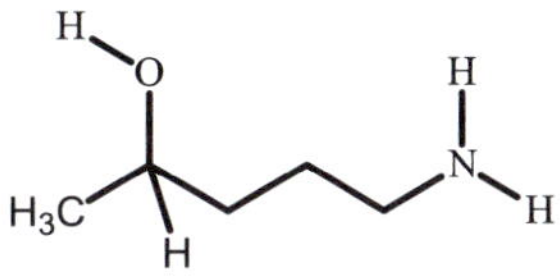

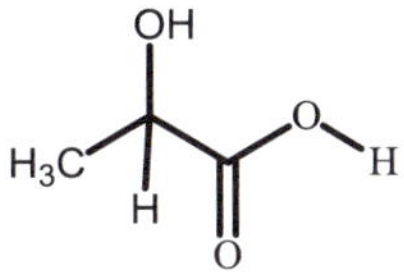

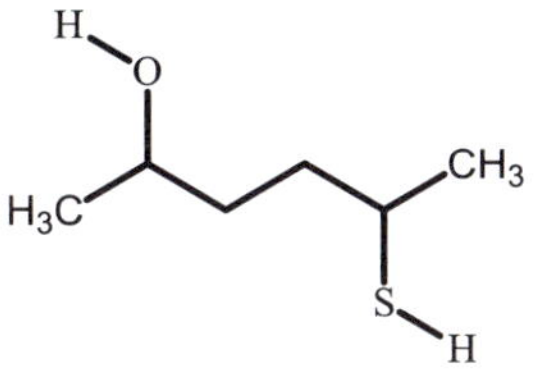

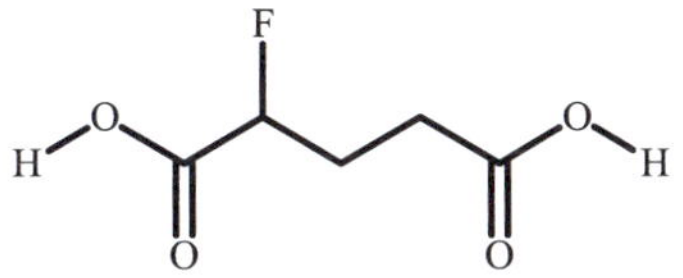

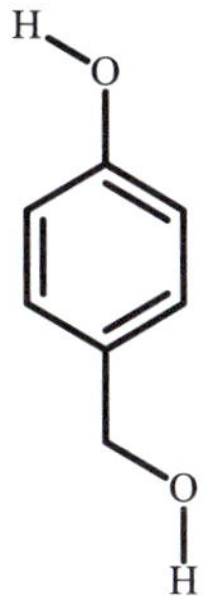

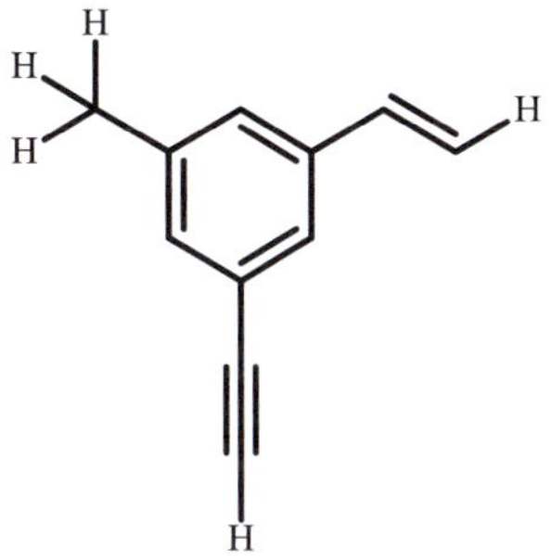

Notes

Lecture Topic I.11: Stability of Cations and Radicals
Hydrocarbon Substituents Stabilize Carbocations

(A) Observation

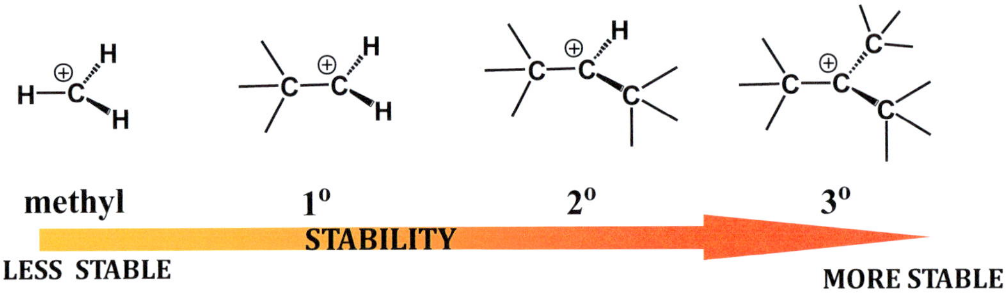

(B) Explanation: Hyperconjugation

Notes

Lecture Topic I.11: Stability of Cations and Radicals
Resonance Stabilizes Carbocations

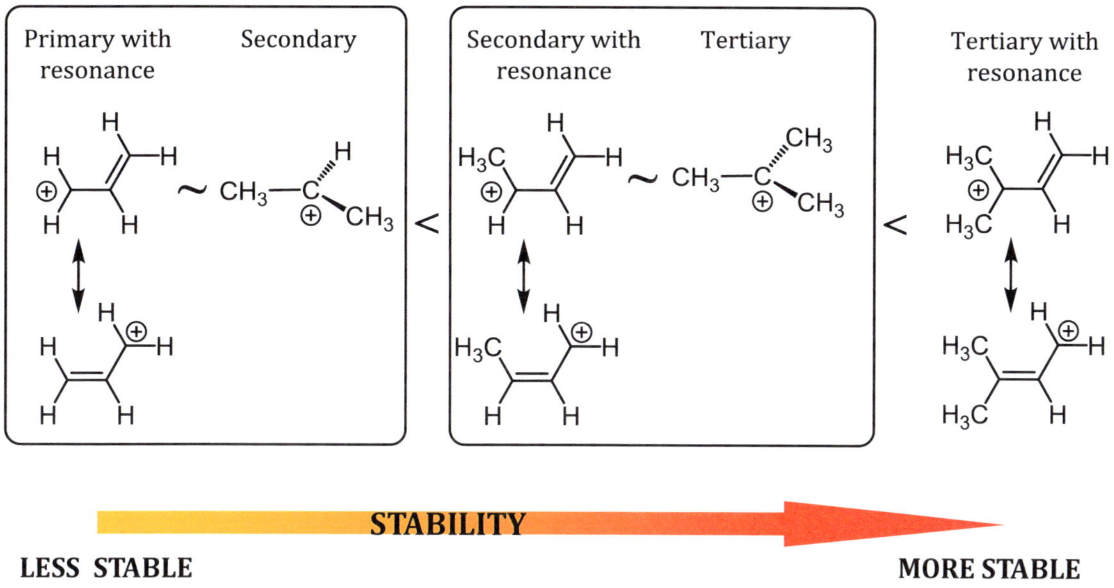

Notes

Lecture Topic I.11: Stability of Cations and Radicals
Substituents Stabilize Carbon-Centered Radicals

Ⓐ Observation

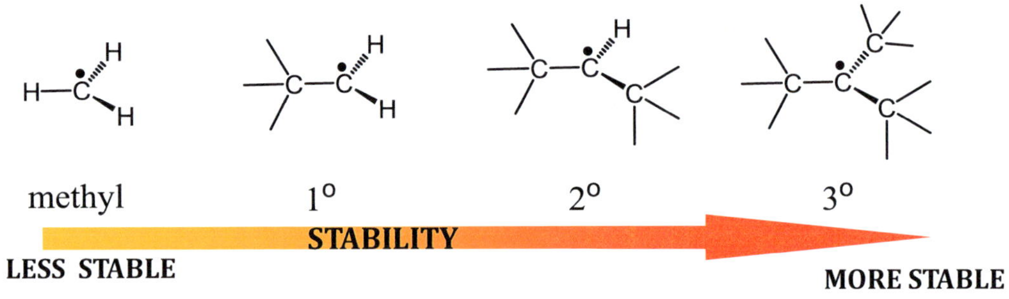

Ⓑ Explanation

Notes

Lecture Topic I.11: Stability of Cations and Radicals
Resonance Stabilizes Carbon-Centered Radicals

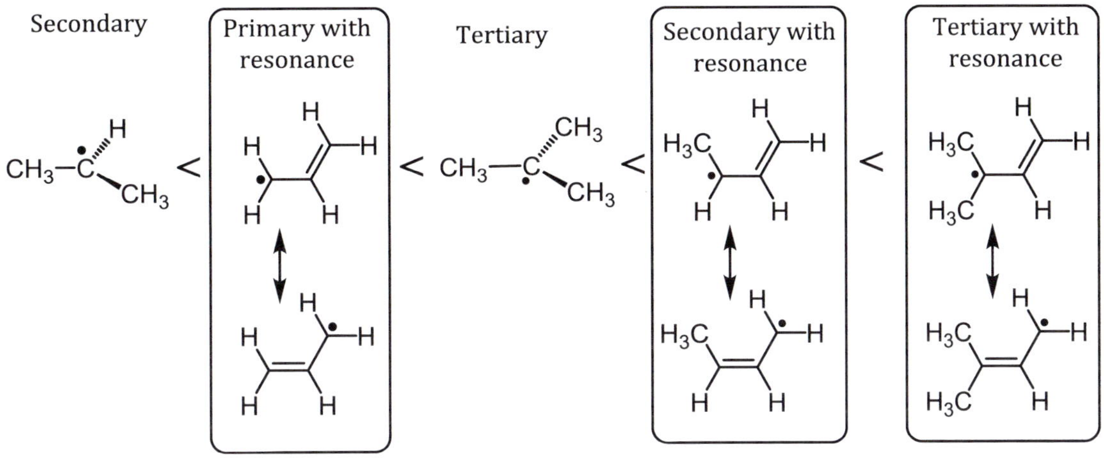

Notes

Lecture Topic I.12: Predicting Reaction Spontaneity and Direction of Equilibria

Spontaneity

Thermodynamically Favorable (Spontaneous):

Example: Assuming that each process is mechanistically accessible, use your knowledge of stability trends to predict whether formation of product would be thermodynamically favorable (spontaneous):

A)

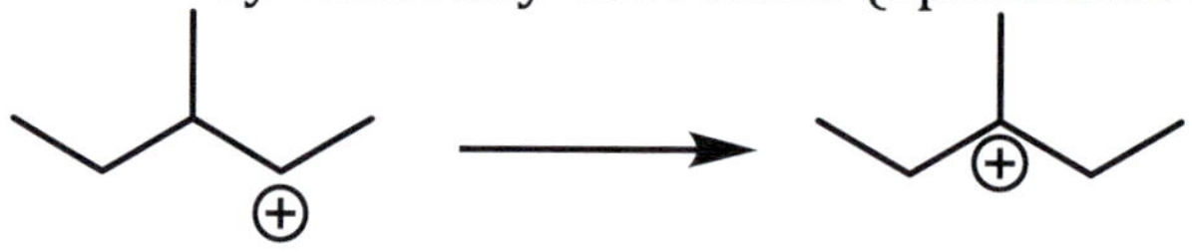

B)

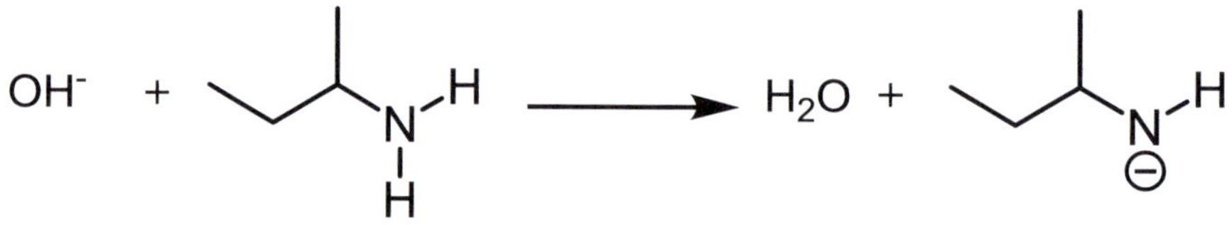

C)

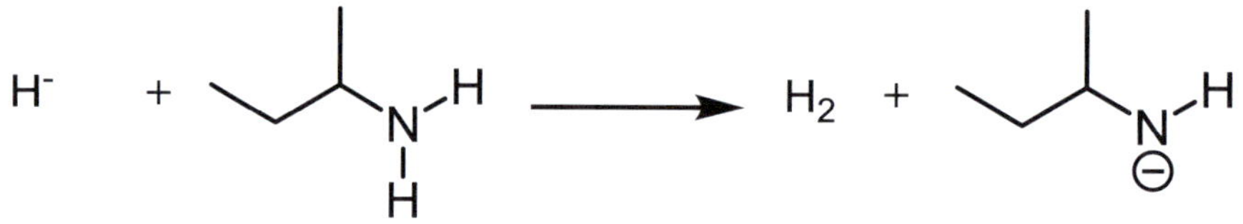

Notes

Lecture Topic I.12: Predicting Reaction Spontaneity and Direction of Equilibria

Predicting Direction of Equilibria

Definition of Equilibrium:

Equations:

Nucleophilic addition example :

S_N2 example:

Nucleophilic Elimination example:

Notes

Lecture Topic I.13: Reaction Coordinate Diagrams and Reaction Rate

Interpreting a Reaction Coordinate Diagram

Stability

Spontaneity:

Reaction Rate:

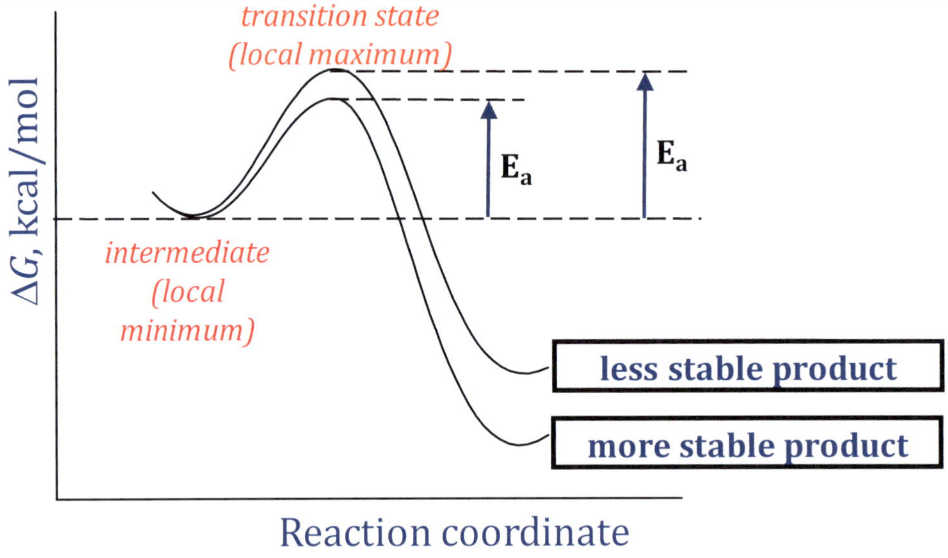

Notes

Lecture Topic I.14: Nomenclature of Alkanes

Linear Alkanes

CH_4		
CH_3CH_3		
$CH_3CH_2CH_3$		
$CH_3(CH_2)_2CH_3$		
$CH_3(CH_2)_3CH_3$		
$CH_3(CH_2)_4CH_3$		
n-C_7H_{16}		
n-C_8H_{18}		
n-C_9H_{20}		
n-$C_{10}H_{22}$		

Notes

Lecture Topic I.14: Nomenclature of Alkanes
Constitutional Isomers and Linear vs. Branched Alkanes

Constitutional Isomers-

C_4H_{10}

linear alkanes
('*n*ormal' alkanes)

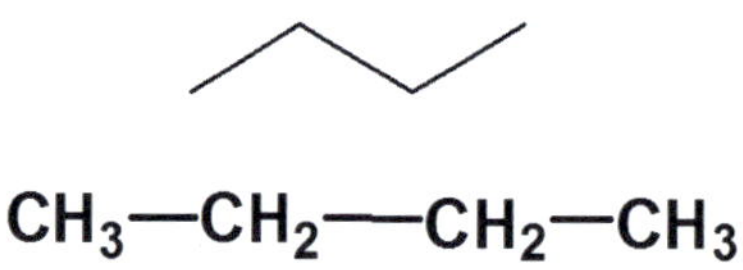

branched alkanes

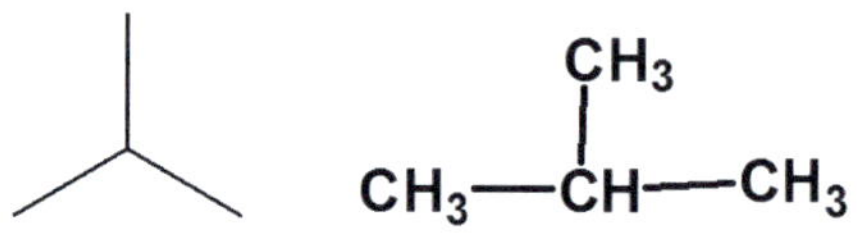

Lighter fluid is 5% *n*-butane and 95% isobutane

Notes

Lecture Topic I.14: Nomenclature of Alkanes

Constitutional Isomers and Linear vs. Branched Alkanes

C_5H_{12}

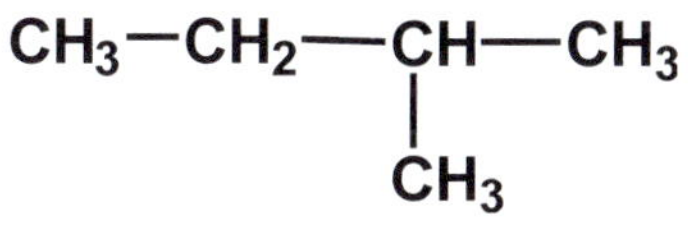

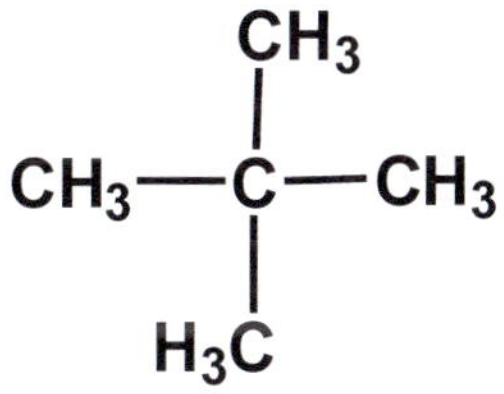

Notes

Lecture Topic I.14: Nomenclature of Alkanes
Stability of Linear vs. Branched Alkanes

Heats of combustion (ΔH_c) for burning C_8H_{18} isomers:

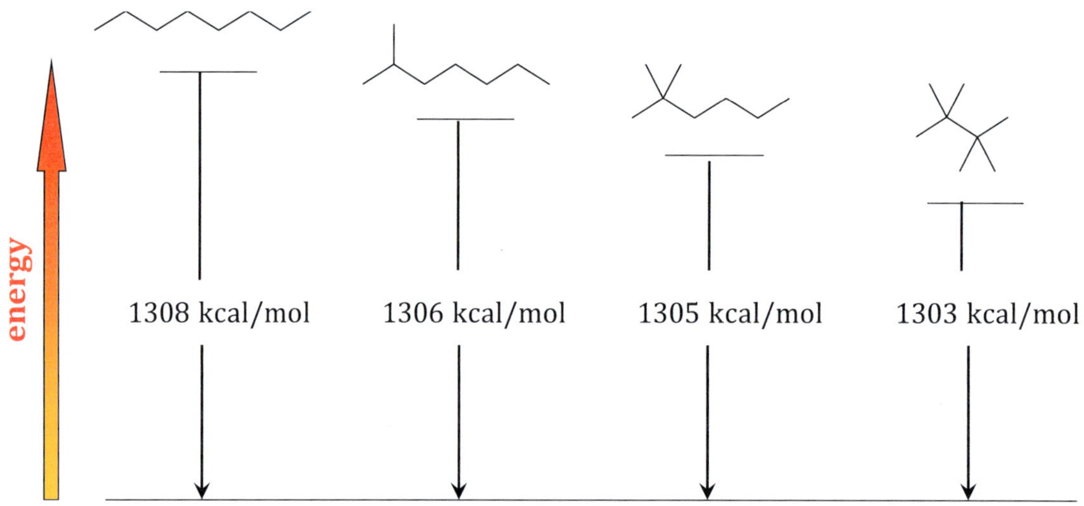

The **more branching**, the **less heat released**, so
more branched =

Notes

Lecture Topic I.14: Nomenclature of Alkanes
Boiling Points of Linear vs. Branched Alkanes

Easier to boil

alkane	bp
CH_4	-160ºC
C_2H_6	-89
C_3H_8	-42
n-C_4H_{10}	-0.4
n-C_5H_{10}	36

Compare linear to branched alkanes:

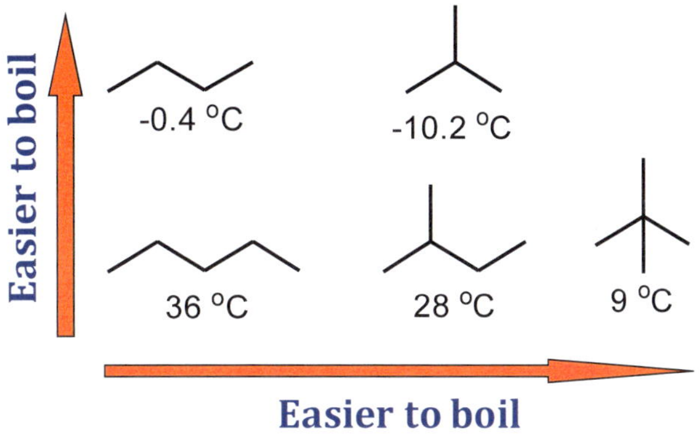

The **lighter** or **more branching**, the **less intermolecular force**, so

more branched =

Lighter =

Notes

Lecture Topic I.14: Nomenclature of Alkanes

Substituents can Modify a Parent Alkane

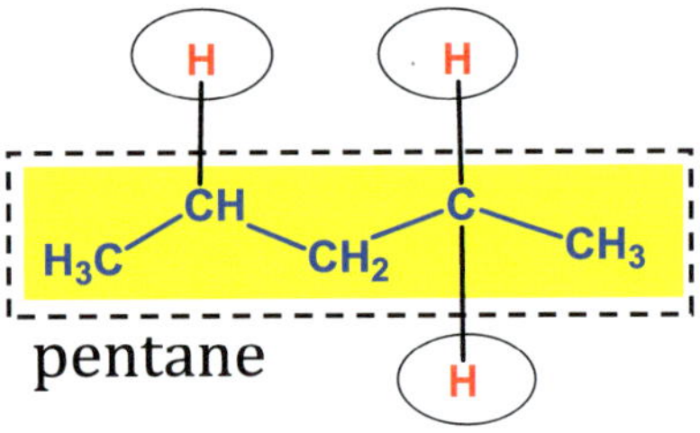

pentane

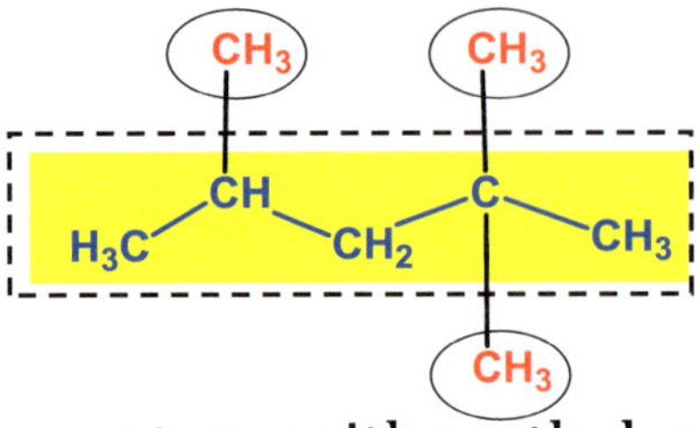

pentane with methyl groups substituted for some of the hydrogens

Substituent:

Substituents that occur frequently in molecules of interest are given names to simplify naming a complex molecule.

Of course there are substituents other than methyl groups we can put on to a molecule ...

Notes

Lecture Topic I.14: Nomenclature of Alkanes
Naming Linear Alkyl Substituents

meth**ane**	CH_4	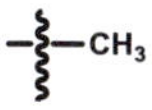	meth**yl**
eth**ane**	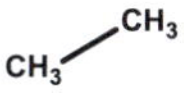	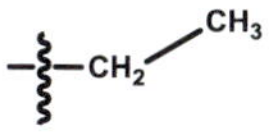	eth**yl**
prop**ane**			n-prop**yl**
n-but**ane**			n-but**yl**
n-pent**ane**			n-pent**yl**
n-hex**ane**			n-hex**yl**
n-hept**ane**			n-hept**yl**
n-oct**ane**			n-oct**yl**
n-non**ane**			n-non**yl**
n-dec**ane**			n-dec**yl**

Notes

Lecture Topic I.14: Nomenclature of Alkanes
Naming Common Non-linear Alkyl Substituents

propane			*n*-propyl	"*n*-" indicates it is "normal", linear
propane			isopropyl	"*iso*-" indicates it is branched
n-butane			*n*-butyl	"*n*-" indicates it is "normal", linear
n-butane		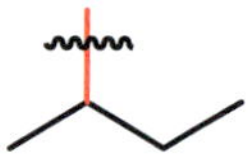	*sec*-butyl	"*sec*-" indicates it is substituted at the secondary carbon
isobutane		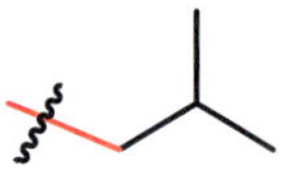	isobutyl	if you substitute the end H, just change "-ane" to "-yl"
isobutane		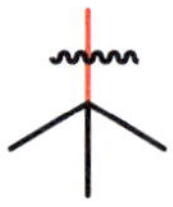	*tert*-butyl	"*tert*-" indicates substitution at the tertiary carbon

Notes

Lecture Topic I.14: Nomenclature of Alkanes
Identifying the Parent Chain

Your step-by-step guide

1.

This is the 'parent chain'; the other stuff on the parent chain will be named as substituents. Be aware that there are many ways to draw the same molecule:

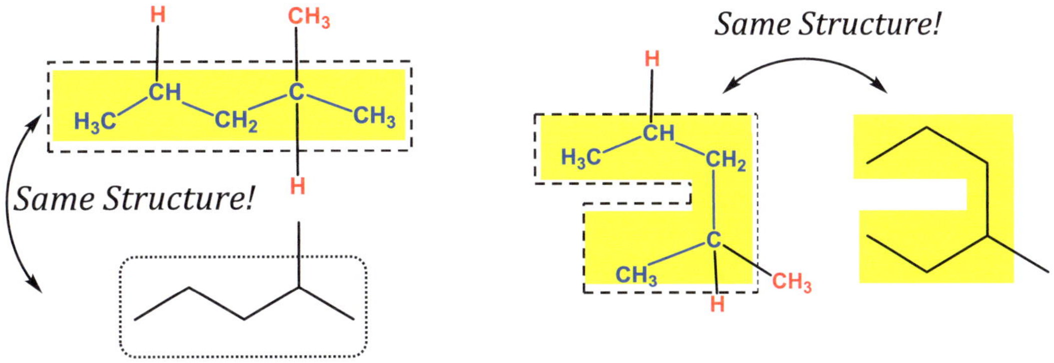

This is pentane, with one methyl (–CH_3) substituent ... regardless of how it is drawn!

Notes

2. Number the carbon atoms in the 'parent chain' in the way that gives the lowest number to the substituent closest to an end of the parent chain.

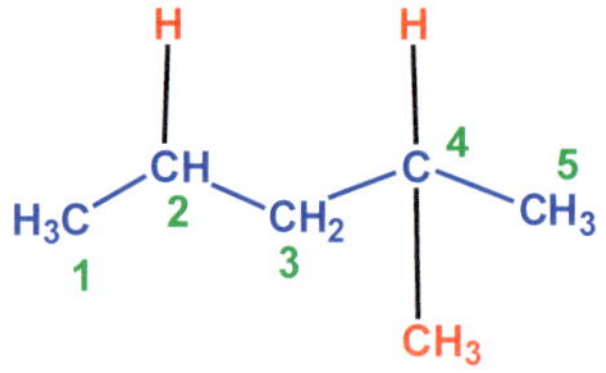

or

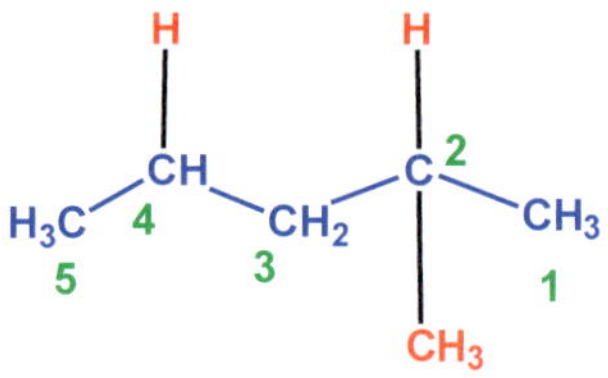

?

(A)

(B)

(C)

This molecule is properly named

<u>Notes</u>

1. Find the longest chain

alkane **possible parent chains:**

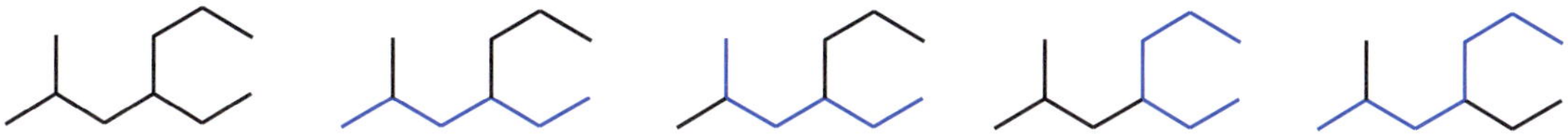

#C's: 6 6 6 7

2. Number the parent chain to give lowest number to substituent closest to end

possible numbering schemes:

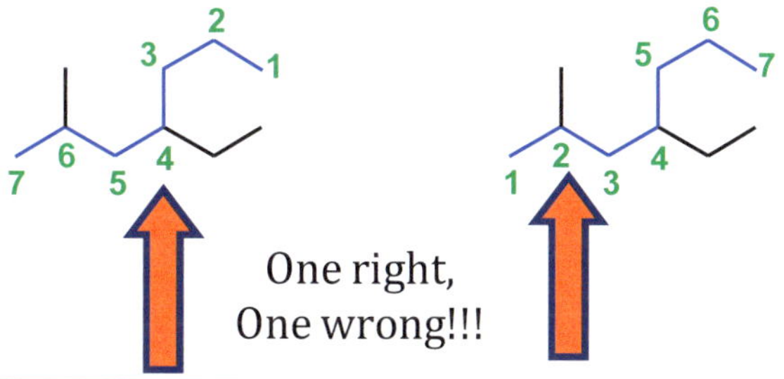

Ⓐ

Ⓑ

Notes

Lecture Topic I.14: Nomenclature of Alkanes

Substituents are Listed Alphabetically in the Name

3. With more than one type of substituent,

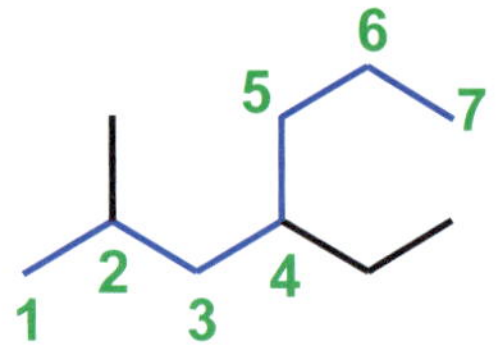

We could name it
2-methyl-4-ethylheptane
or
4-ethyl-2-methylheptane

Since 'ethyl' is alphabetically before 'methyl', this molecule is properly named

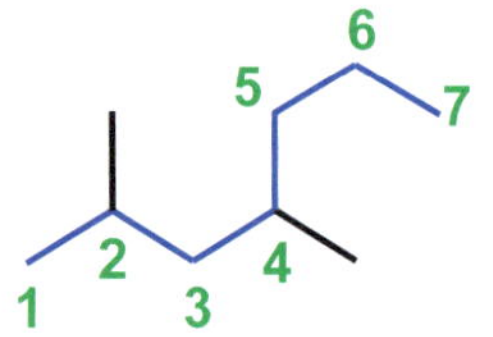

Now consider this closely related molecule, in which the ethyl group is changed to a methyl group. The numbering and parent chain are the same, but now we have two methyl groups; we need to apply the next rule ...

Notes

4. If more than one of the same substituent are present on your parent chain, use di, tri, tetra, etc. prefixes to denote this (**these prefixes do not count when alphabetizing.** Neither do the *n*-, *sec*-, or *tert*- prefixes.

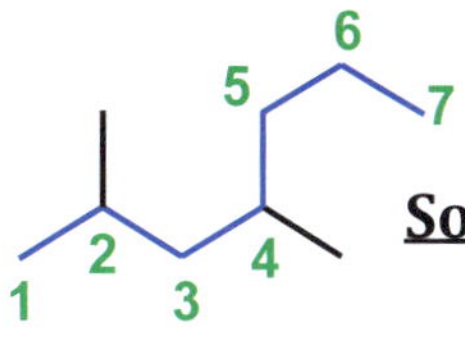

So, we call this

~~NOT~~
~~2-methyl-4-methylheptane~~
~~or~~
~~4-methyl-2-methylheptane~~

Incorrect names!

Notes

Lecture Topic I.14: Nomenclature of Alkanes
Alphabetically First Substituents Get the Lower Number in Case of a Tie

There are even more potential complications that could trip us up; consider:

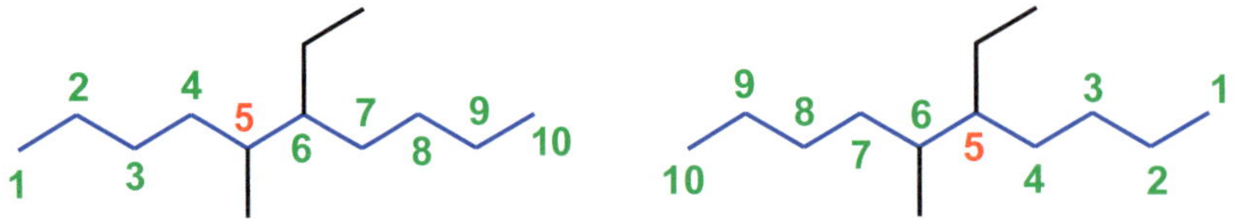

We've easily found the parent chain (decane), and we try to find the right numbering scheme; but either way we count, the first substituent is at the 5 position; which is right?
We need the next rule:

5. If numbering leads to the same "lowest number" substituent in both directions the correct numbering:

In this case, ethyl is alphabetically before methyl, so ethyl is given the 5:

This molecule is properly named:

Notes

Lecture Topic I.14: Nomenclature of Alkanes
With two Possible Parent Chains, Choose the More Substituted

The previous example showed you how to deal with substituent numbering issues; but what if you have trouble identifying the parent chain to begin with? Consider:

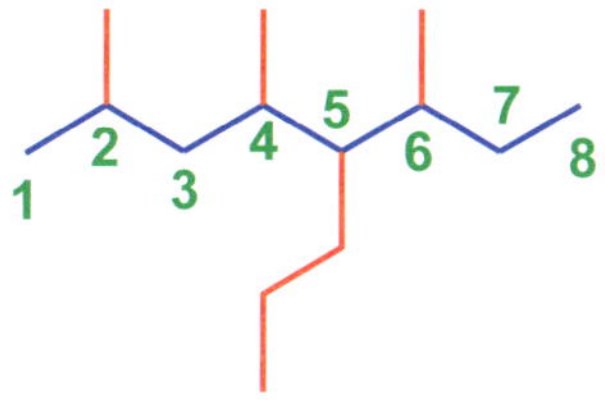

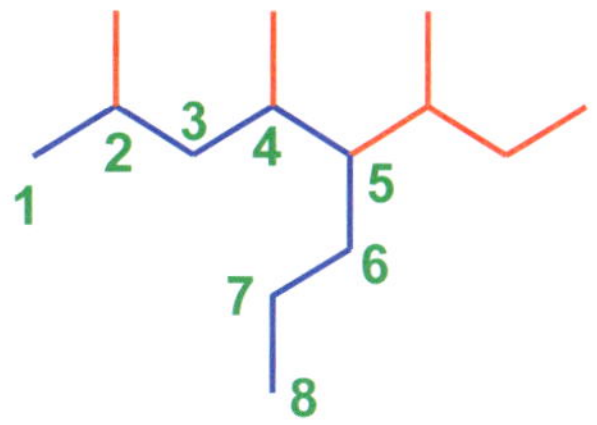

T**wo** possible octane parent chains; which is right? We need another rule.

6. If you find two different possible parent chains of the same length,

The one on the left has more substituents, so we use that parent chain.

This molecule is properly named:

Notes

Lecture Topic I.14: Nomenclature of Alkanes

Summary of Rules for Naming Simple Alkanes

Your step-by-step guide to alkane nomenclature

1. Find the longest chain. This is the 'parent chain'; the other stuff on the parent chain will be named as substituents. Be aware that there are many ways to draw the same molecule.

2. Number the carbon atoms in the 'parent chain' in the way that gives the lowest number to the substituent closest to an end of the parent chain.

3. With more than one type of substituent, name in alphabetical order.

4. If more than one of the same substituent are present on your parent chain, use di, tri, tetra, etc. prefixes to denote this (these prefixes do not count when alphabetizing, though; neither do the n-, sec-, or tert- prefixes)

5. If numbering leads to the same lowest number substituent in both directions the correct numbering gives the lowest number to the substituent that is first alphabetically.

6. If you find two different possible parent chains of the same length, you choose the one with more substituents coming off of it.

Notes

Lecture Topic I.15: Cycloalkanes and Configurational Isomerism

Cycloalkanes

Cyclic versions of alkanes are called cycloalkanes, and are named by the prefix used in the linear alkane with the same number of carbon atoms:

A
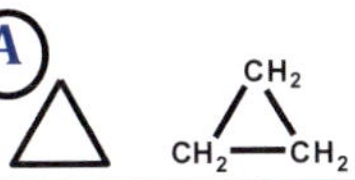

B
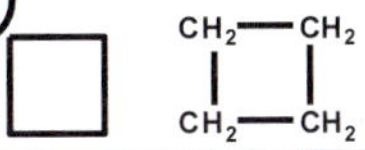

C
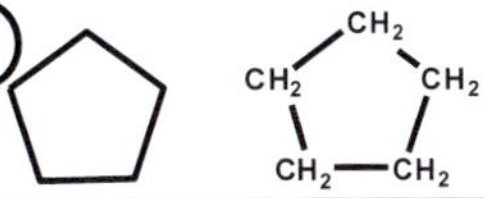

D
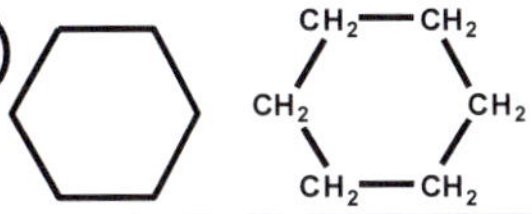

E
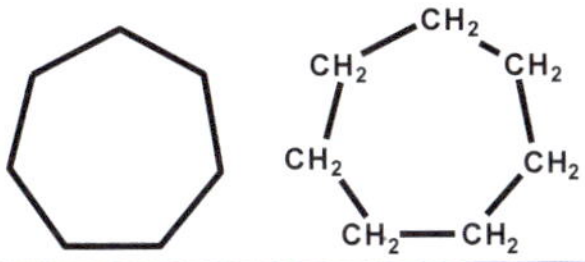

F
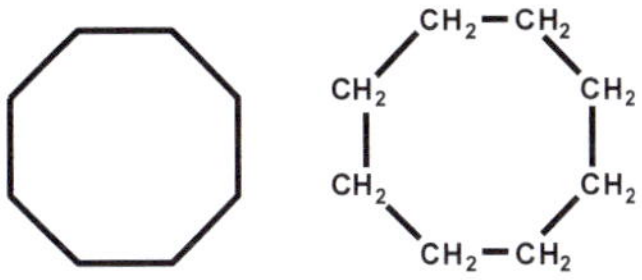

Of course, cycloalkanes can have substituents as well, and we need to know how to name substituted cycloalkanes.

Notes

Lecture Topic I.15: Cycloalkanes and Configurational Isomerism
Naming Substituted Cycloalkanes

Armed with our knowledge of naming branched alkanes, naming cycloalkanes is pretty straightforward. We can use the cycloalkane as the parent chain (if it is longest) and follow numbering/alphabetizing rules from there.

An additional rule for naming cycloalkanes:
If there is a possible parent chain on the cycloalkane that is longer than the # of carbons in the cycloalkane, use that as the parent chain, and name the cycloalkane as a substituent. For example:

(A) 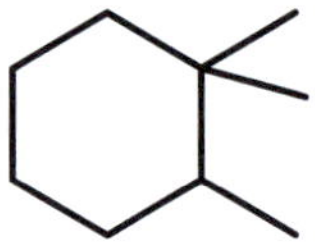

(B) 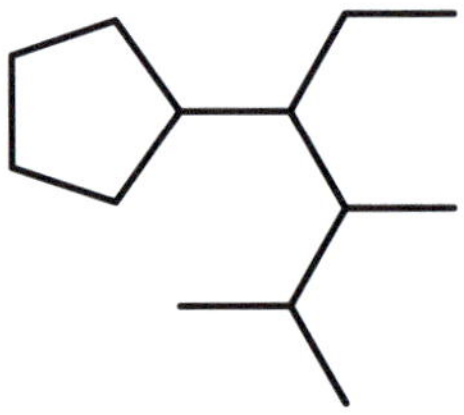

Notes

Depending on where we put the methyl substituents, we have two different **configurational isomers**.

A configurational isomer in which substituents are on the same side (both 'up' in this case) is referred to as a ***cis-*** isomer, while a configurational isomer with substituents on opposite sides is referred to as the ***trans-*** isomer.

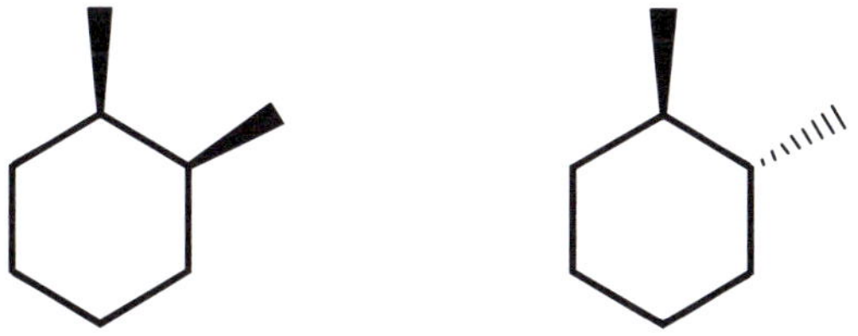

Notes

Lecture Topic I.16: Nomenclature of Alkyl Halides and Alcohols

Naming Alkyl Halides

For alkyl halides, name the compound using the rules for naming alkanes and cycloalkanes, and treat the halides as substituents:

CCl_2H_2 dichloromethane

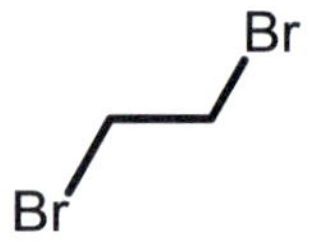

1,2-dibromoethane

-F = Fluoro
-Cl = Chloro-
-Br = Bromo-
-I = Iodo-

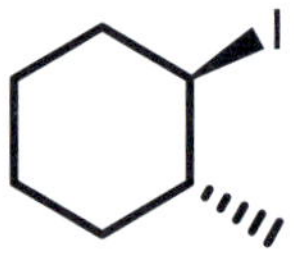

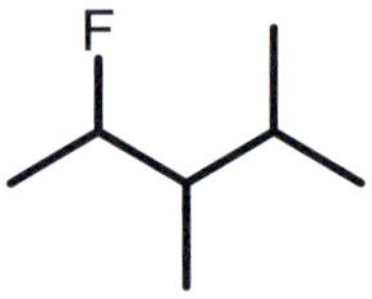

__Notes__

Lecture Topic I.16: Nomenclature of Alkyl Halides and Alcohols
Naming Alcohols

For alcohols, use the rules for naming alkanes and cycloalkanes as a starting point, with the following adjustments:

1. replace the "e" at the end of the alkane name with "ol".

2. The alcohol is always given the lowest possible number. Note that this means that the alcohol is always given the "1" position in cycloalkanes (so, no need to add a number; we know it's always 1).

3. Place the number indicating the position of the alcohol directly before the parent chain name (which now ends in "ol").

<u>*Notes*</u>

Lecture Topic I.16: Nomenclature of Alkyl Halides and Alcohols
Naming Alcohols

CH_3OH Methanol

CH_3CH_2OH Ethanol

1-propanol
(or *n*-propanol)

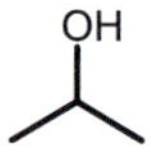

2-propanol
(or isopropanol)

A

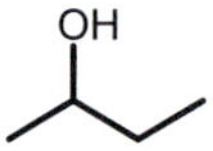

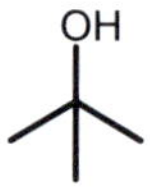

2-methyl-2-propanol
(or *tert*-butanol)

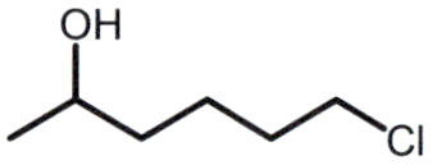

6-chloro-2-hexanol

B

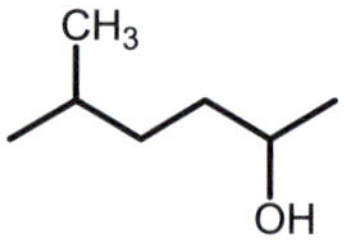

trans-4-methylcyclohexanol

C

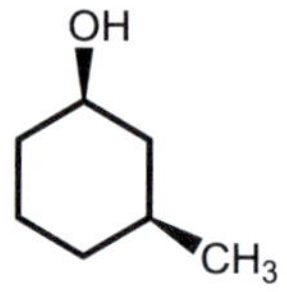

Notes

Lecture Topic I.16: Nomenclature of Alkyl Halides and Alcohols
Classifying Alcohols and Alkyl Halides

We learned to classify carbocations and radicals as methyl, primary (1°), etc. We can classify alcohols/alkyl halides similarly.

$HO{-}CH_2R$

primary alcohol (1° alcohol)

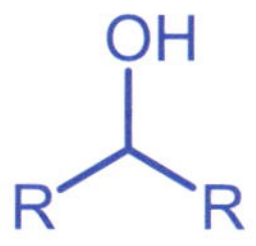

secondary alcohol (2° alcohol)

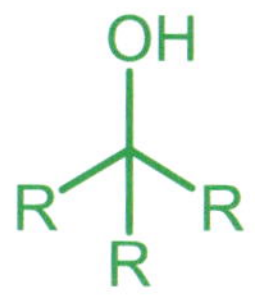

tertiary alcohol (3° alcohol)

$X{-}CH_2R$

primary alkyl halide (1°)

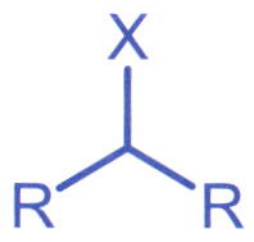

secondary alkyl halide (2°)

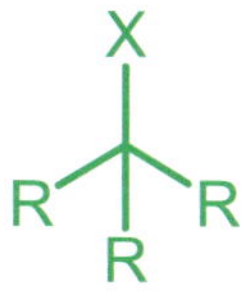

tertiary alkyl halide (3°)

These classifications are very important, because the different classes of alcohols and alkyl halides have different reactivities, as we will learn in detail later in the course.

Notes

Lecture Topic I.17: Isomerism and Conformational Analysis I

Rotating around a C–C Bond Leads to Different Conformations

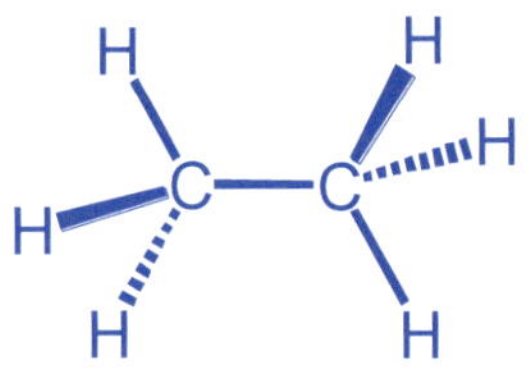

In ethane, two tetreahedral C are bonded to one another. We can think of this geometrically by joining two tetrahedra at a corner:

the eye sees:

The two triangles are

the eye sees:

The two triangles are

There are different **conformations** in which the two tetrahedra can be oriented with respect to one another, based on the relative rotation of the tetrahedral C atoms about the C–C bond.

Notes

Lecture Topic I.17: Isomerism and Conformational Analysis I

Rotation about a C–C Single Bond Leads to Conformational Isomers

The use of wedges and hashed lines lets us represent the possible 3D shapes so we can visualize the possible **conformations**:

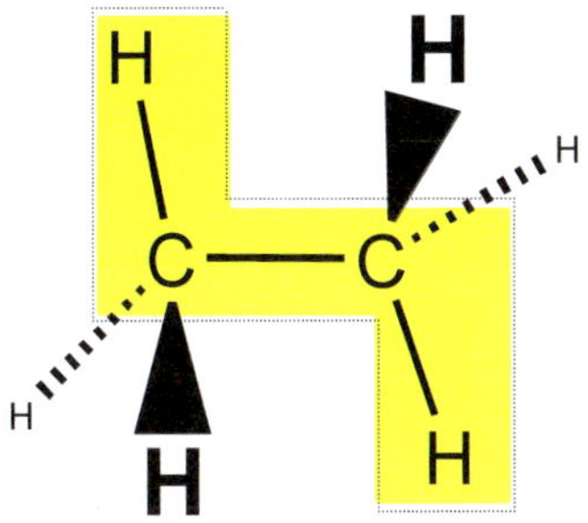

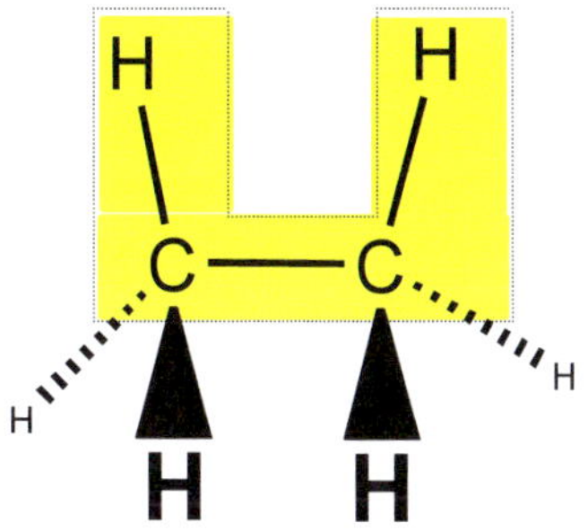

Notice in the left conformer the two H's coming out at us are far apart from one another, while in the right conformer they are right next to each other. These are two **conformational isomers -**

Notes

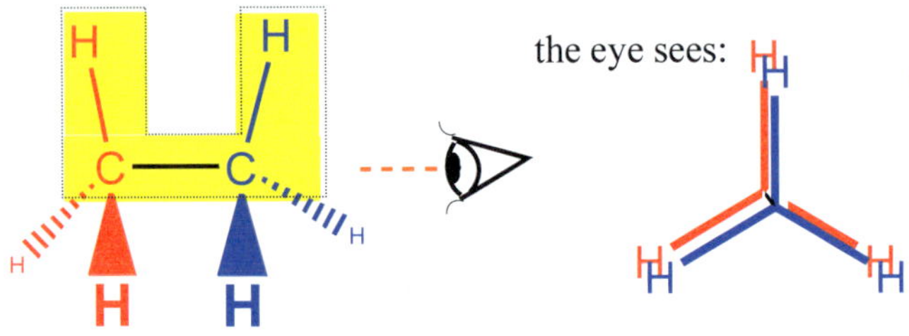

(A) This conformer of ethane is called the

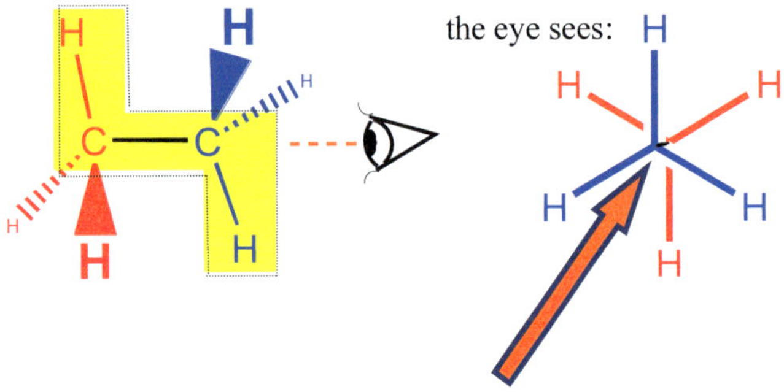

(B) This conformer of ethane is called the

These representations emphasize the fact that we have two different conformers, but are a bit messy, especially at the middle, where all the bonds converge. For this reason, another way of representing this has been developed ...

Notes

Put a disk in to block the view of the middle portion of the back, so that the middle is more tidy, and you get what are called

(A)

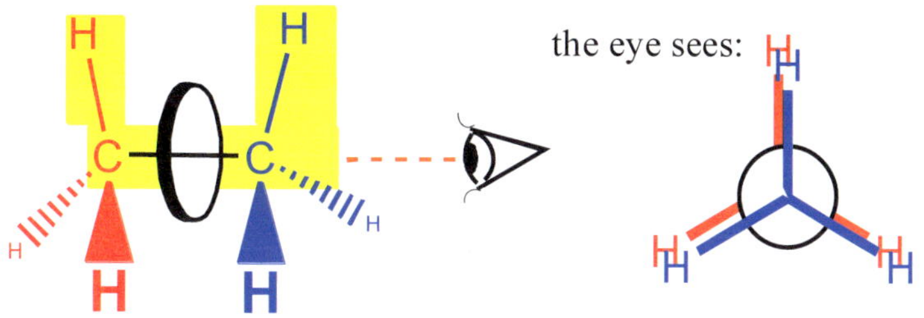

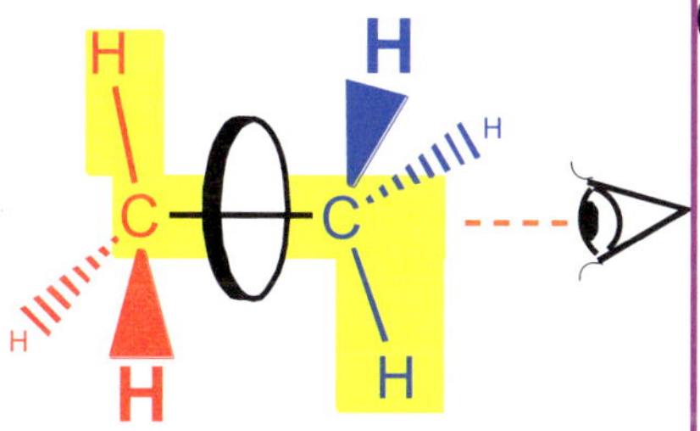

(B)

If you cut a little circle of paper, put in on your model kit ethane, and look from the end, you'll see exactly what the Newman Projection represents.

Notes

Steric Repulsion Destabilizes Eclipsed Conformations

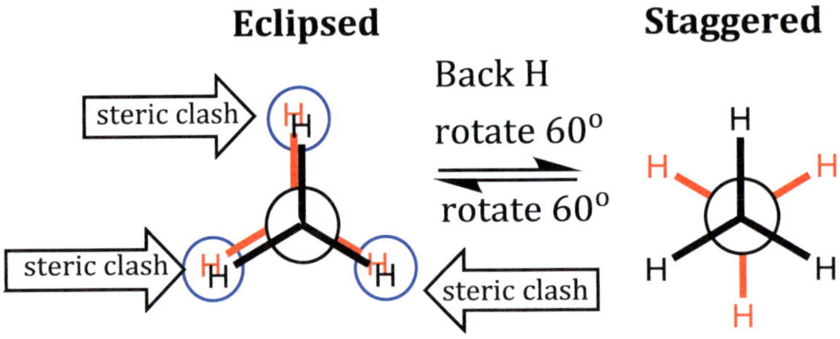

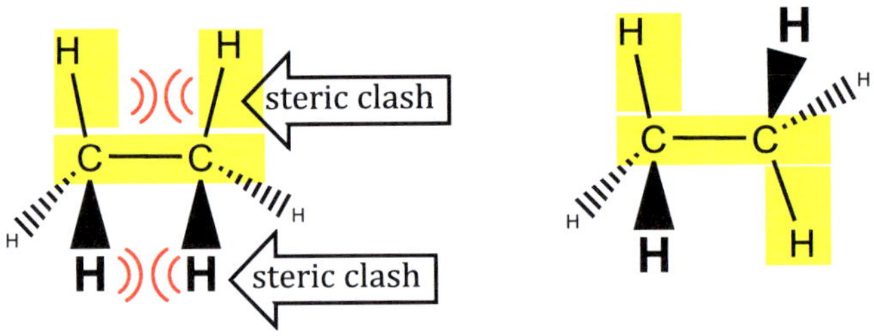

In the eclipsed conformation, H substituents are physically bumping into one another; such interactions are called **steric** interactions (steric interactions lead to different stereoisomers). In staggered, the H atoms are far apart. Because of sterics,

Notes

Steric clashes leads to an energy barrier (E_a) to rotation about the C-C bond. This energy barrier is called the **torsional strain** (resistance to twisting). Below is an energy profile showing how stability changes with bond rotation.

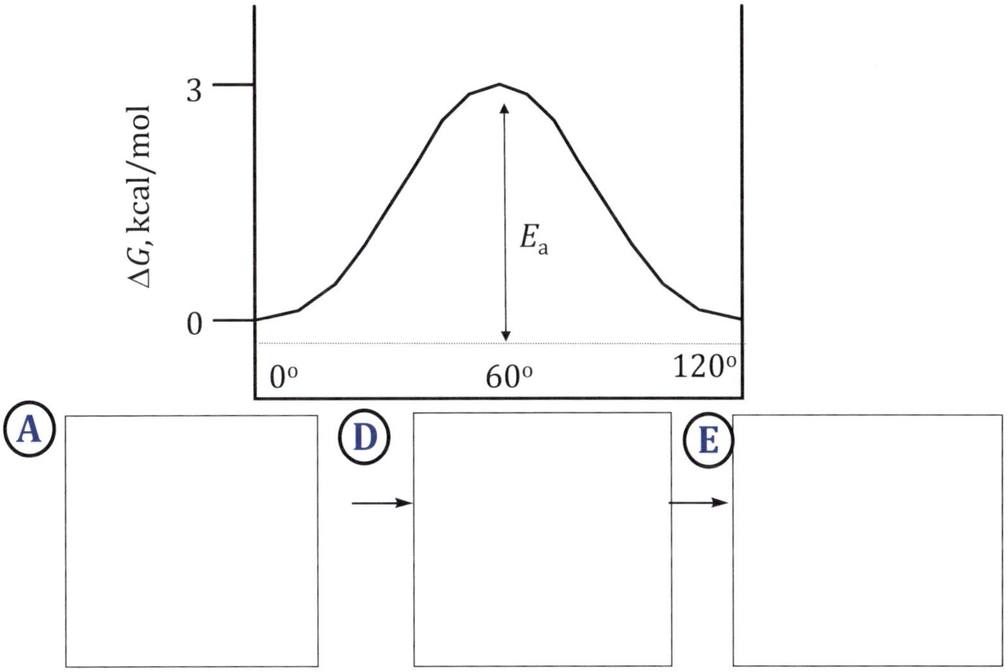

Notes

Lecture Topic I.17: Isomerism and Conformational Analysis I
Concepts to Analyze Ethane can be applied to other Molecules

Ethane is simple because we only have hydrogens on the C-C bond. Now we look at a more complicated example, butane.

$H_3C—CH_3$

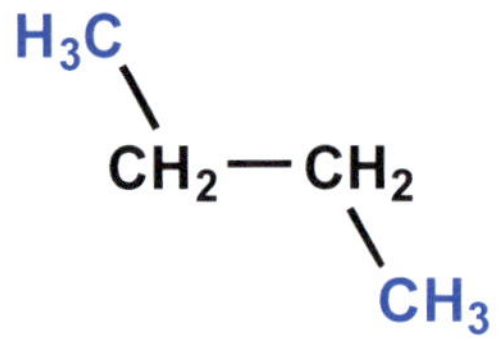

Butane has both hydrogen and methyl groups attached to the C2-C3 bond. This leads to more possible stereoisomers than we saw with ethane ...

Ethane **Butane**

Notes

Lecture Topic I.17: Isomerism and Conformational Analysis I
There are Four Staggered/Eclipsed Conformations of Butane

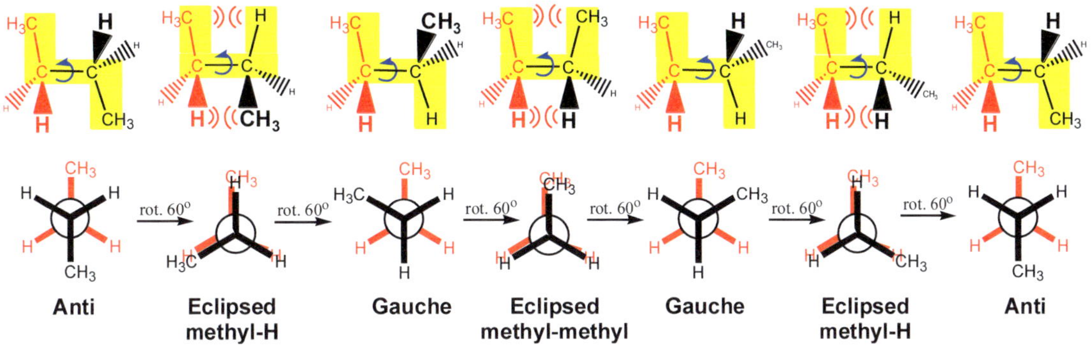

four distinct conformers:

We can predict the relative stability of these conformations by evaluating the steric strain in each of them...

Notes

Lecture Topic I.17: Isomerism and Conformational Analysis I
Assessing Sterics Allows Prediction of Stability

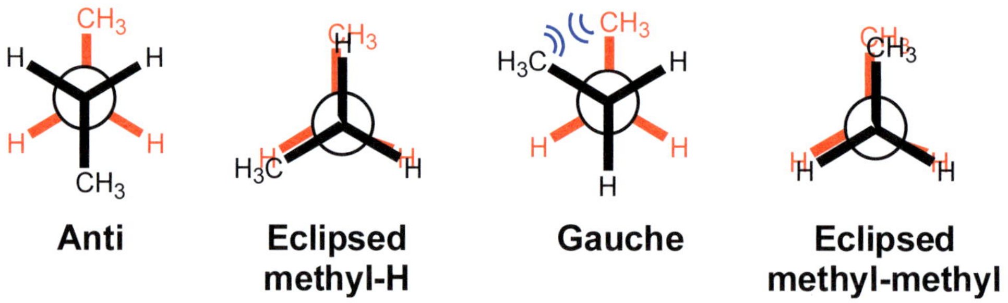

Since methyl groups are larger than hydrogens,

Ⓐ

In fact, two methyl groups have some steric repulsion even when they are still 60° apart from one another, a repulsion called a

Ⓑ

Notes

Lecture Topic I.17: Isomerism and Conformational Analysis I
Assessing Sterics Allows Prediction of Stability

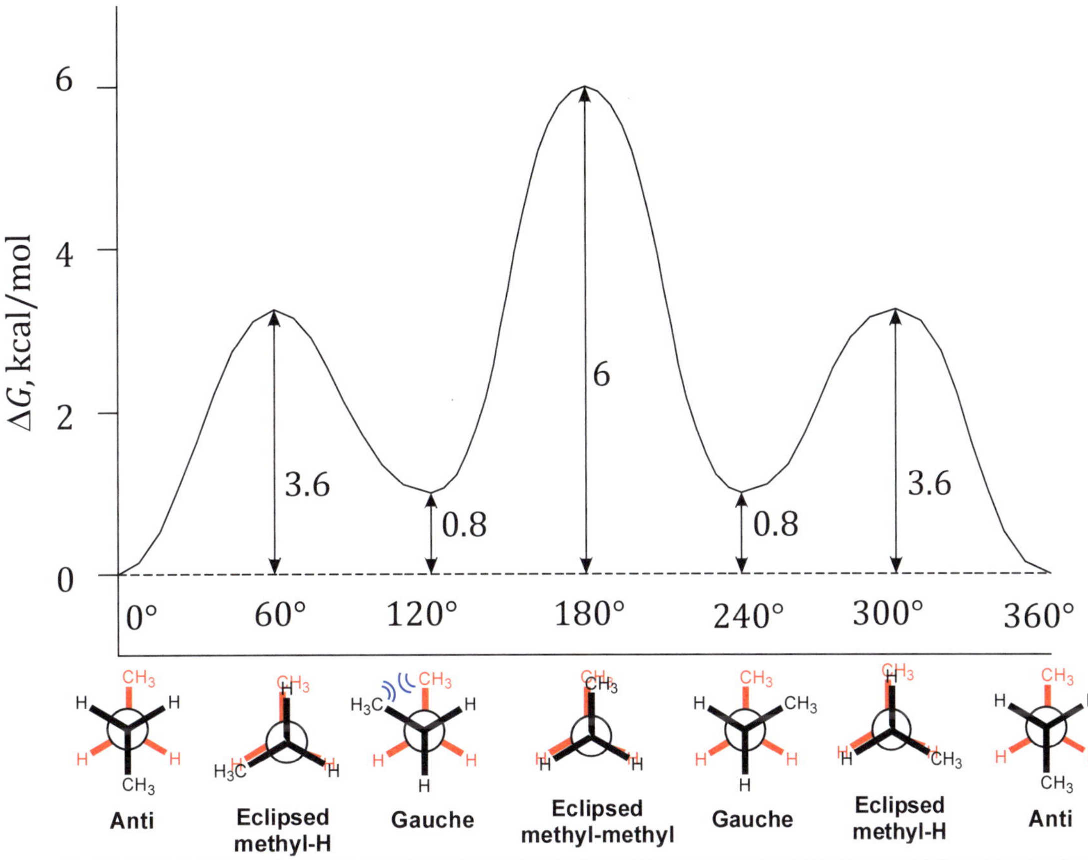

Notes

Lecture Topic I.18: Conformational Analysis II
Strain in Cycloalkanes

We have discussed steric strain (including torsional strain) in linear alkanes.

One additional consideration for cyclic compounds is **angle strain**. Cycloalkanes are made of sp^3 hybridized carbons, which favor bond angles of 109.5°. However, this differs from the angles in some regular polygons:

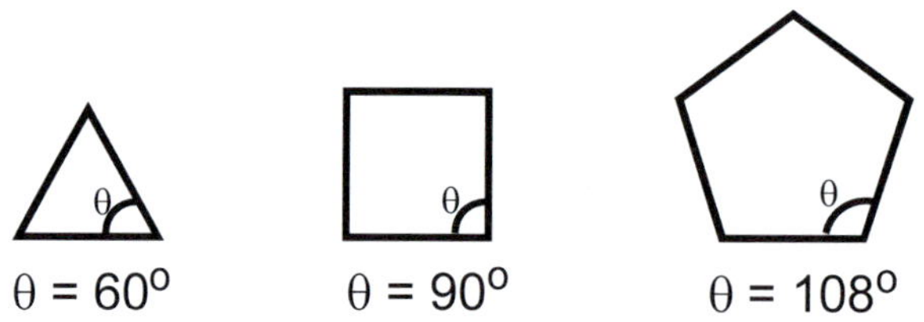

The difference between ideal and actual angle leads to **angle strain**. We consider this as part of the overall **Ring strain:**

Notes

Lecture Topic I.18: Conformational Analysis II

Angle Strain is in part Due to Poor Hybrid Orbital Overlap

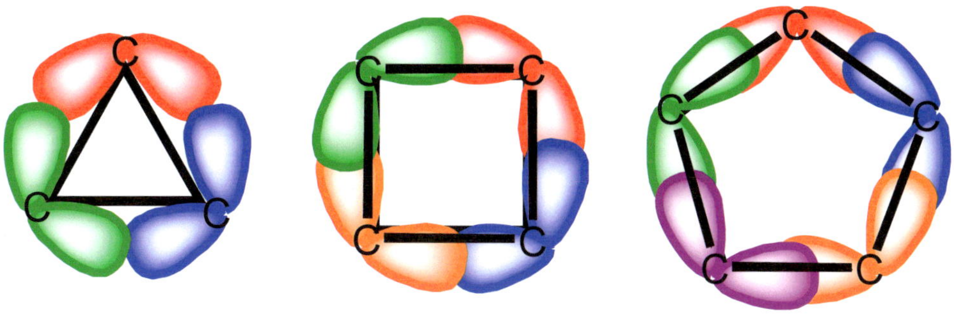

An **origin of angle strain**

Now we consider the other contributors to ring strain: steric interactions (and torsional strain)

Notes

Lecture Topic I.18: Conformational Analysis II
Ring Strain in Cyclopropane

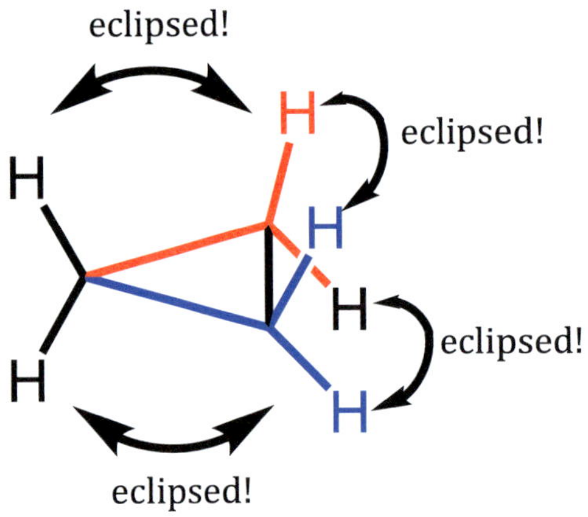

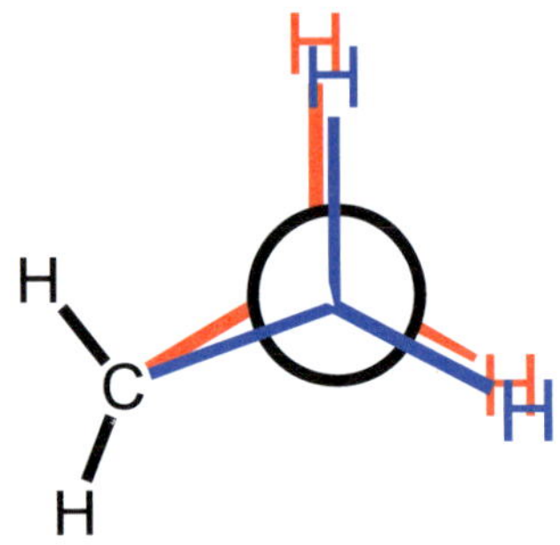

In addition to lots of **angle strain**, cyclopropane also has

The total **ring strain** is about 30 kcal/mol (10 kcal per methylene).

Notes

Lecture Topic I.18: Conformational Analysis II

Ring Strain in Cyclobutane

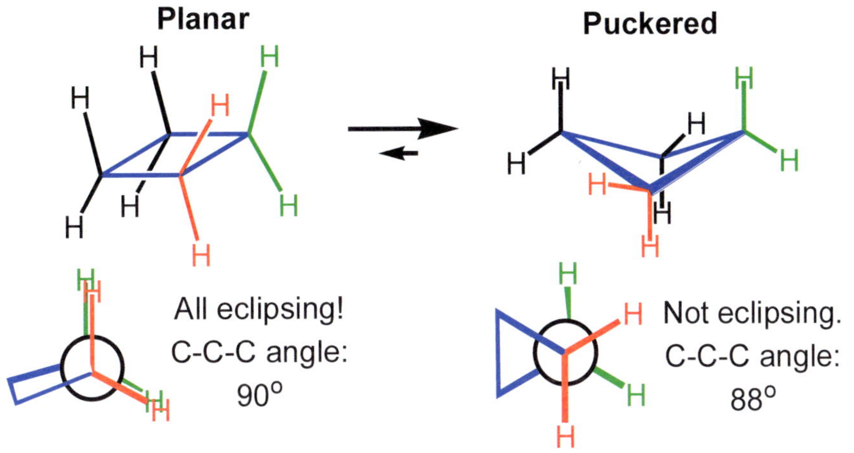

In addition to **angle strain**, planar cyclobutane has H's eclipsed. **Molecules will always**

Cyclobutane twists ("puckers"), increasing the angle strain a bit, but relieving all of the eclipsing torsional strain. The puckered form, on the right, is the favored form. The total **ring strain** is about 28 kcal/mol (7 kcal per methylene).

Notes

Lecture Topic I.18: Conformational Analysis II
Ring Strain in Cyclopentane

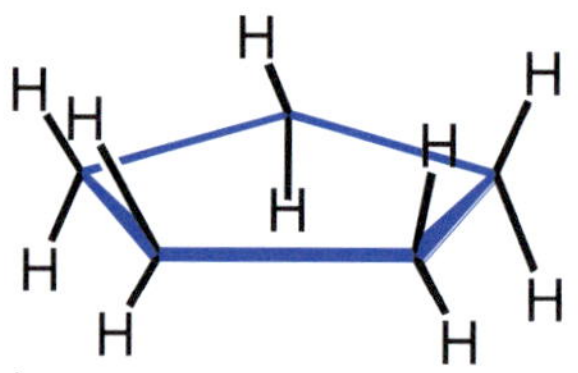

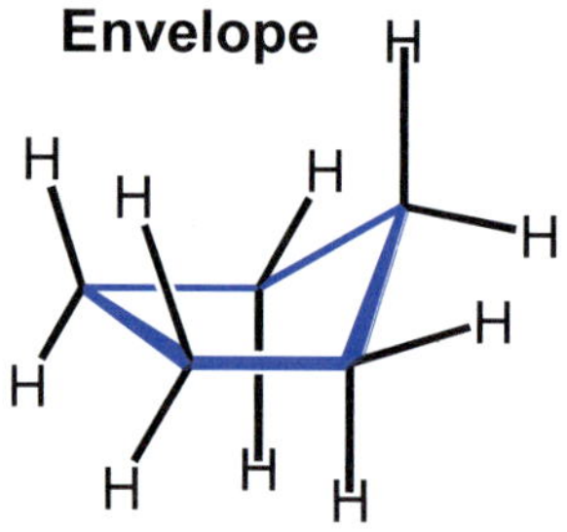

In addition to some **angle strain**, planar cyclopentane also has

(A) [ ]

Cyclopentane undergoes a distortion to the **envelope form**, relieving some of the eclipsing torsional strain, so:

(B) [ ]

The total **ring strain** is about 10 kcal/mol (2 kcal per methylene).

__Notes__

Lecture Topic I.18: Conformational Analysis II
Ring Strain in Cyclohexane

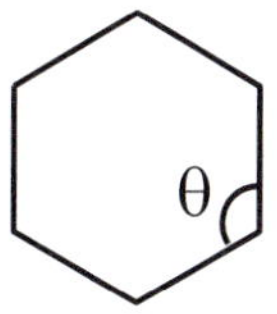

$\theta = 120^{o}$

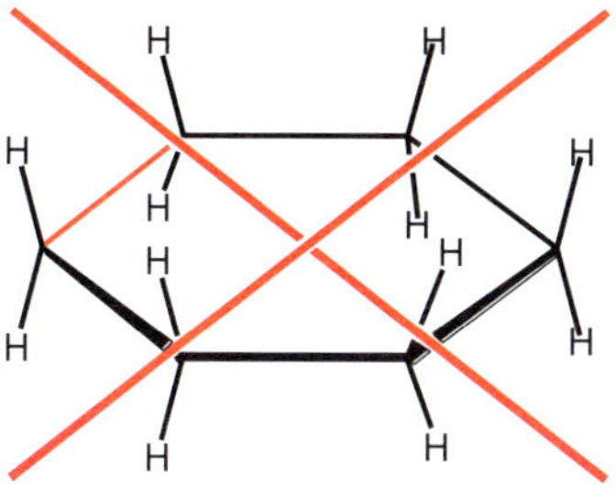

Planar: ring strain from:

(A)

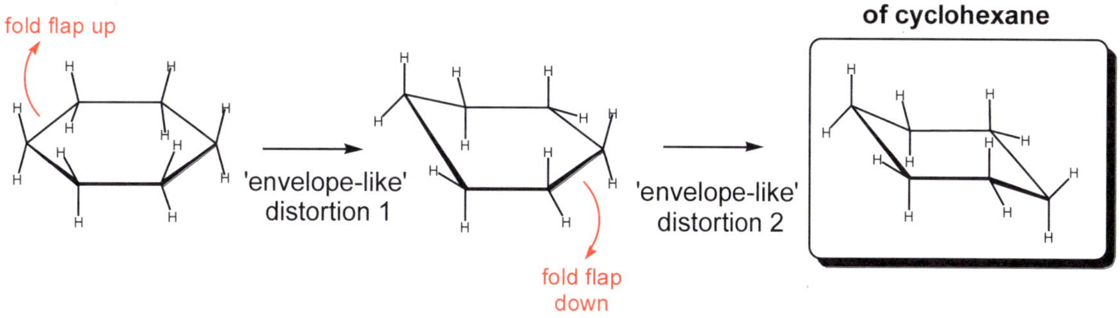

Chair conformation of cyclohexane

Ring strain in the Chair Conformation:

We will now study why this is the case.

Notes

Lecture Topic I.18: Conformational Analysis II
The Chair Conformation of Cyclohexane

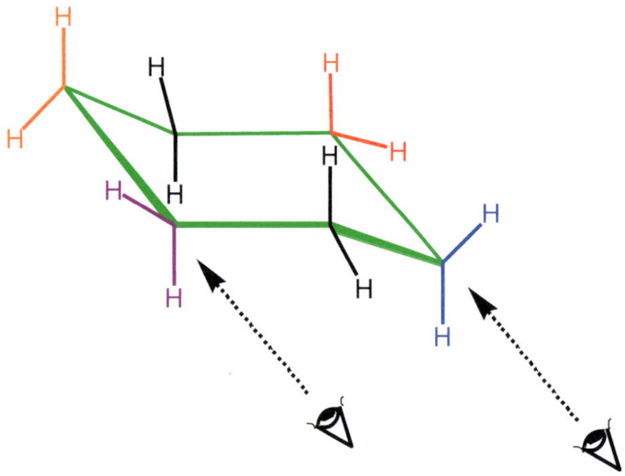

All staggered.

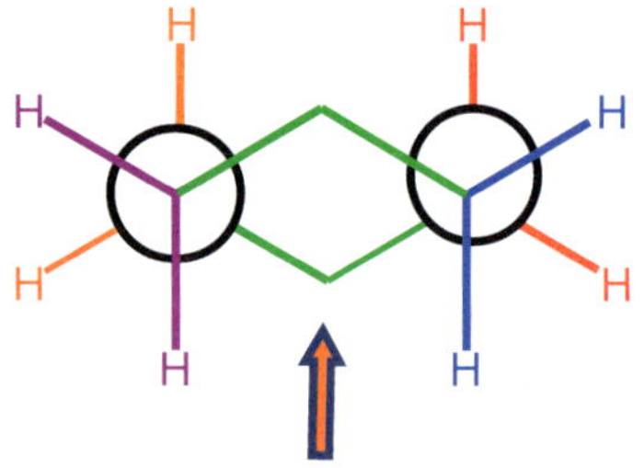

This picture shows two Newman Projections; one projection down the right side and one down the left side of cyclohexane (indicated by the eyes). Build a model to see this more easily.

The chair conformation features
1.

2.

This is why the chair conformation is very stable.

Notes

Lecture Topic I.18: Conformational Analysis II

There are two types of Positions for Substituents on Chair Cyclohexane

There are two distinct types of hydrogen atoms in cyclohexane. It is very important to understand the difference between these two types.

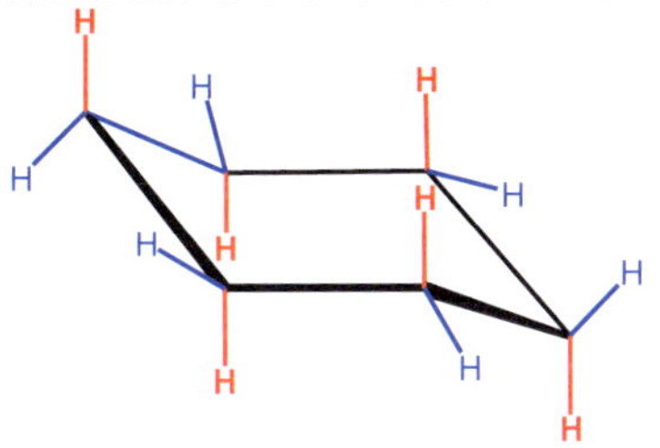

Note: Each C has one equatorial and one axial H and each type alternates up and down around the ring.

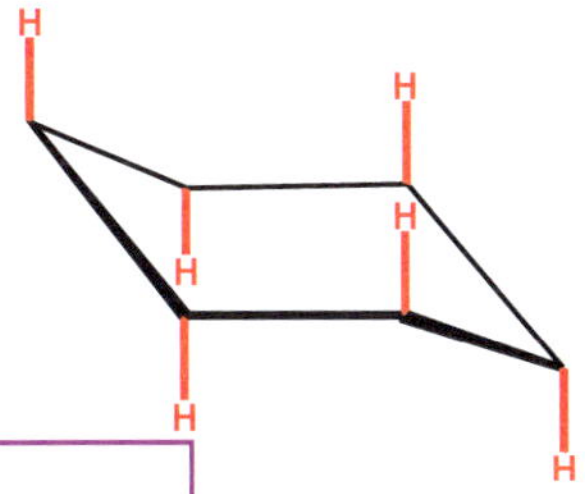

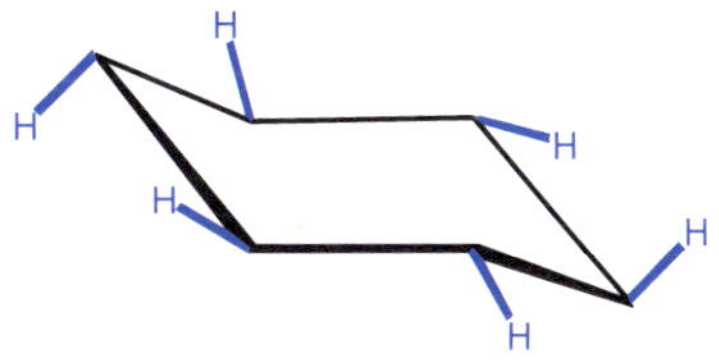

__________: 'stick straight up and down'. If you build a model of the chair conformation of cyclohexane these are the atoms that it will 'stand on.'

__________: point up and down at an angle around the <u>equator</u> of the cyclohexane.

Notes

Lecture Topic I.18: Conformational Analysis II
Ring Flipping in Chair Cyclohexane

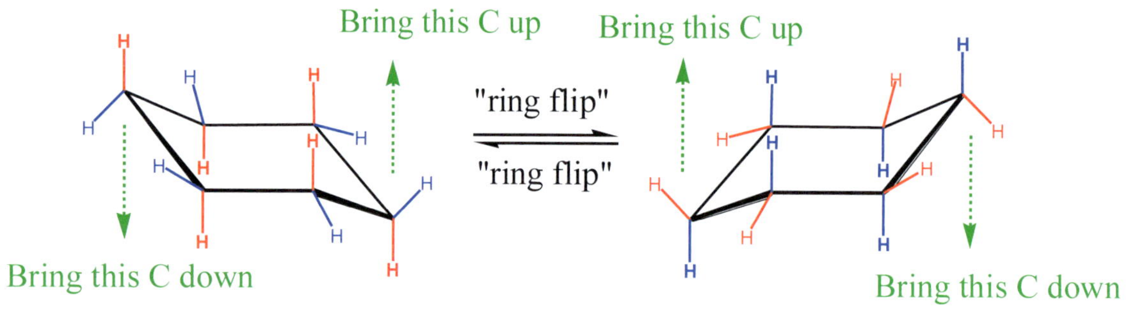

A **ring flip** involves distorting the cyclohexane ring so that the 'up' flap moves to the 'down' position and vice versa.

A **ring flip** switches:

Notes

Lecture Topic I.18: Conformational Analysis II

Steric Repulsion Disfavors Axial Substitution

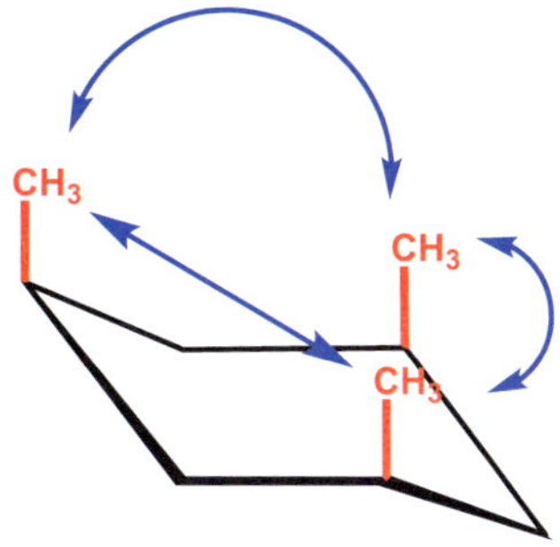

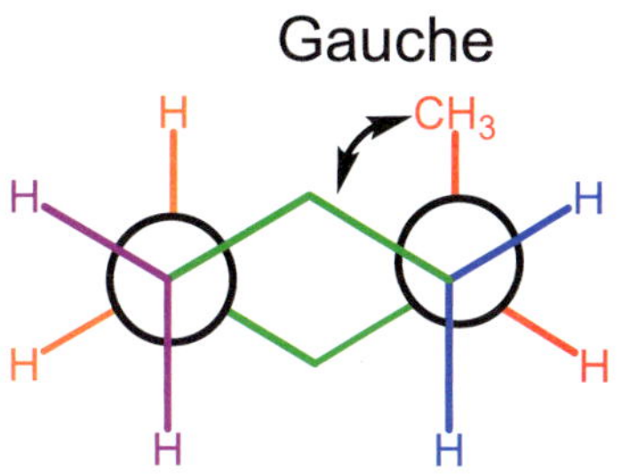

Substituents may cause steric strains. Axial methyl (or larger) groups can clash with other axial substituents on the same face of the ring. This type of steric interaction is called a

(A)

Equatorial positions do not have these interactions, so

(B)

Notes

Lecture Topic I.18: Conformational Analysis II

Lower Steric Strain leads to the Most Stable Conformation

Problem: For each, draw the most stable conformational isomer

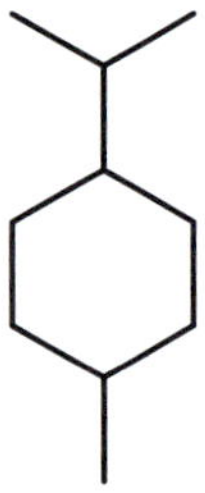

1-isopropyl-4-methylcyclohexane

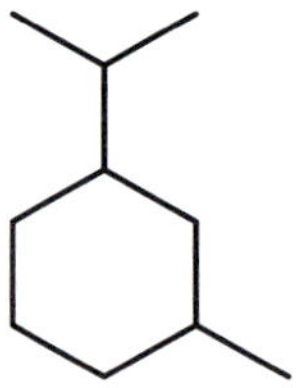

1-isopropyl-3-methylcyclohexane

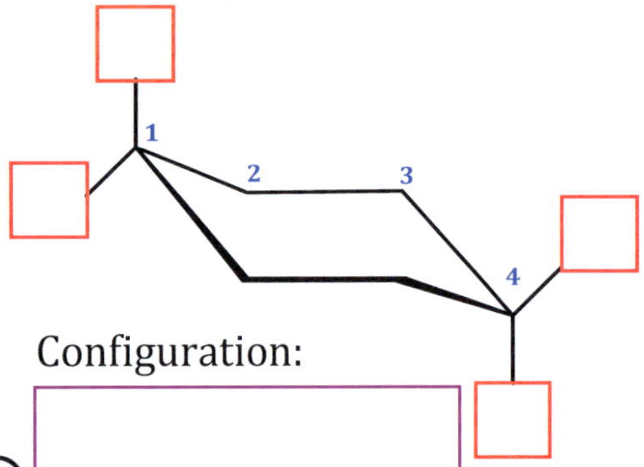

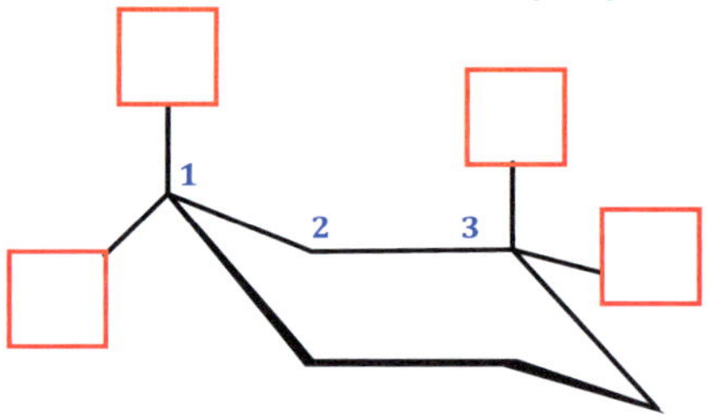

Configuration:

(A)

Configuration:

(B)

Notes

Lecture Topic I.18: Conformational Analysis II

Determining the Most Stable Conformer

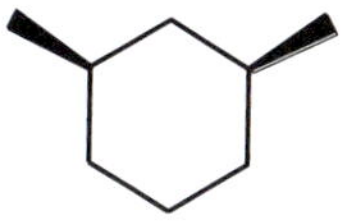

cis-1,3-dimethycyclohexane

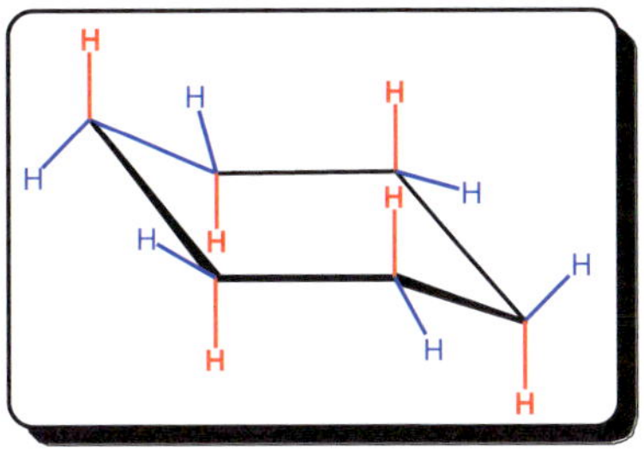

For Reference

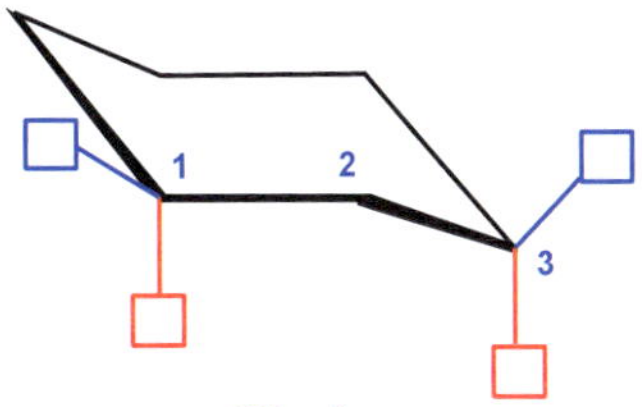

Choices

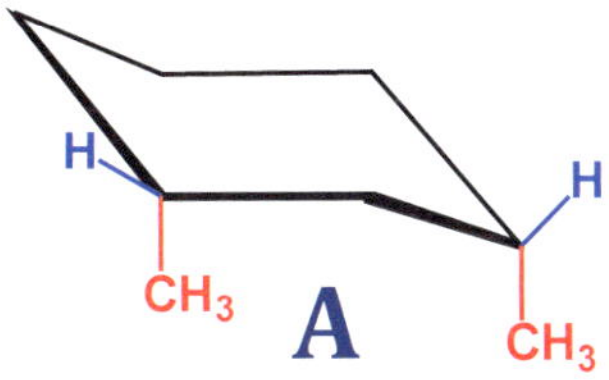

A

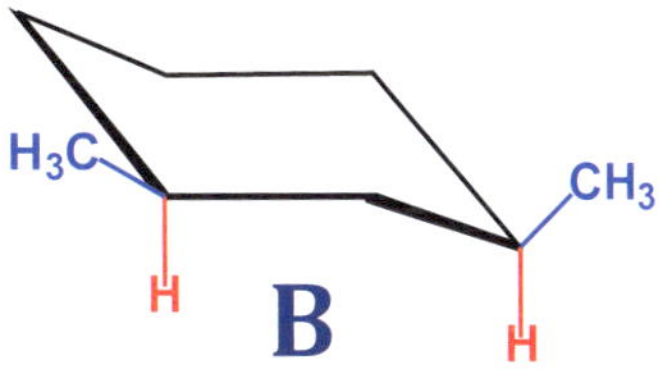

B

Notes

Lecture Topic I.18: Conformational Analysis II
Determining the Most Stable Conformer

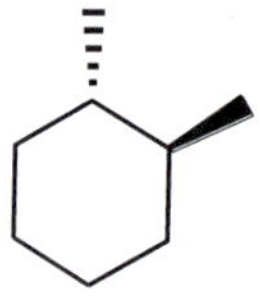

trans-1,2-dimethycyclohexane

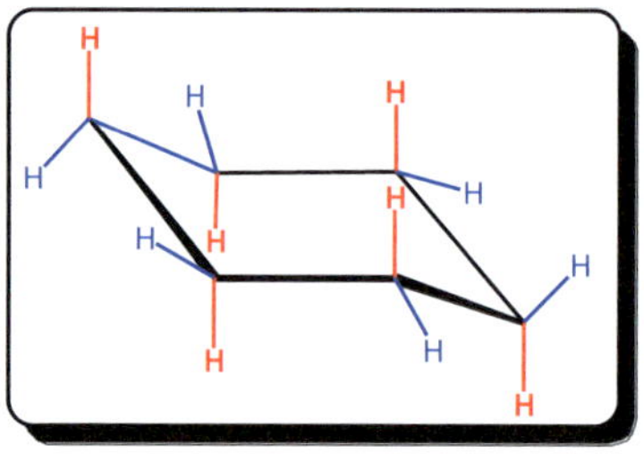

For Reference

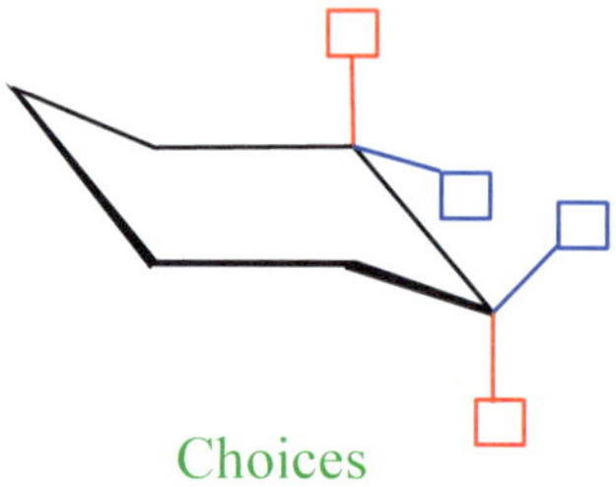

Choices

(A)

(B)

Axial methyls will have strong 1,3-diaxial repulsion, so this form will be unfavorable. The most stable form is therefore:

(C)

Notes

Lecture Topic I.18: Conformational Analysis II
The Boat Conformation of Cyclohexane

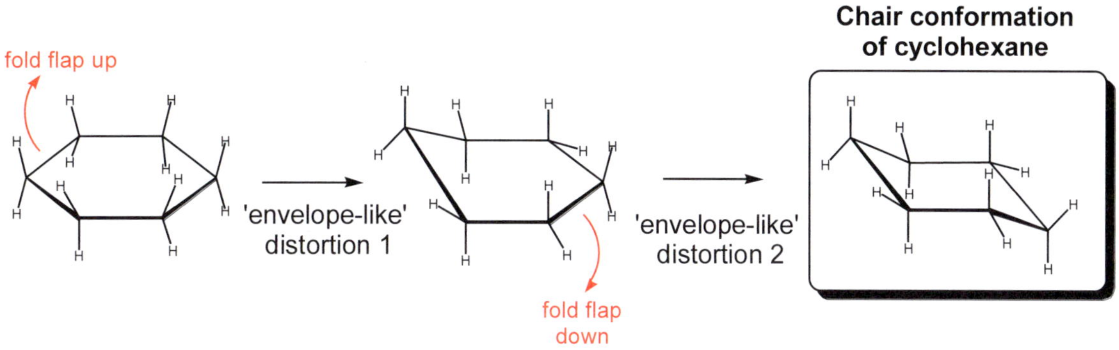

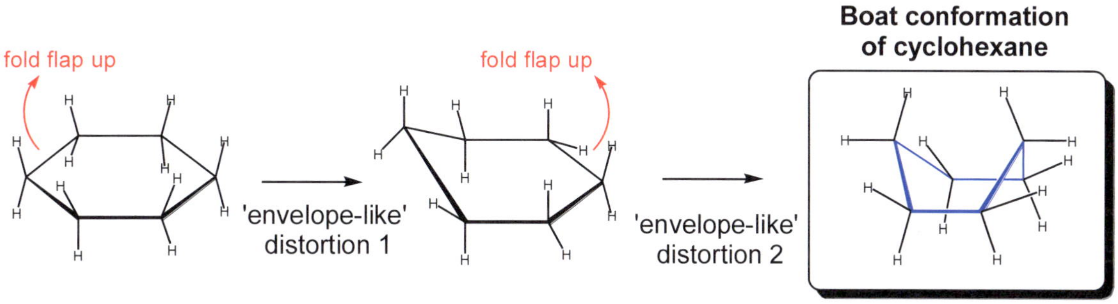

Notes

Lecture Topic I.18: Conformational Analysis II
The Boat Conformation of Cyclohexane

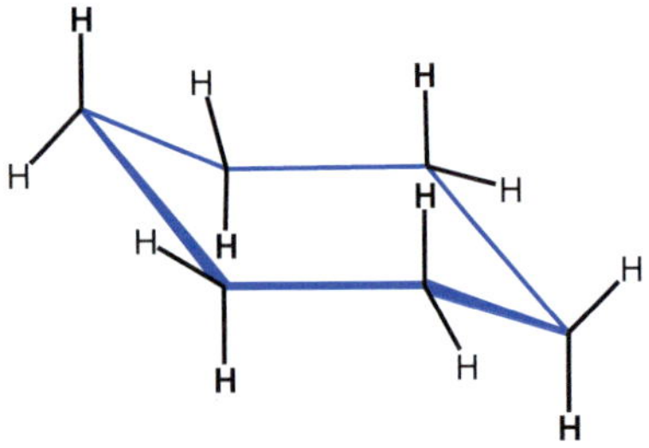

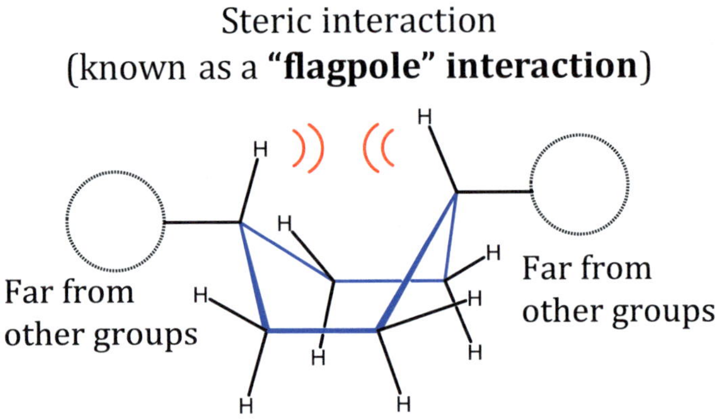

"**chair**" conformation is very stable:
- no eclipsing interactions
- angles favorable
- Generally preferred

"**boat**" conformation
- eclipsing interactions
- steric strain
-angles strained

Boat only preferred when:

Notes

Lecture Topic I.19-22: Stereochemistry
Types of Isomerism

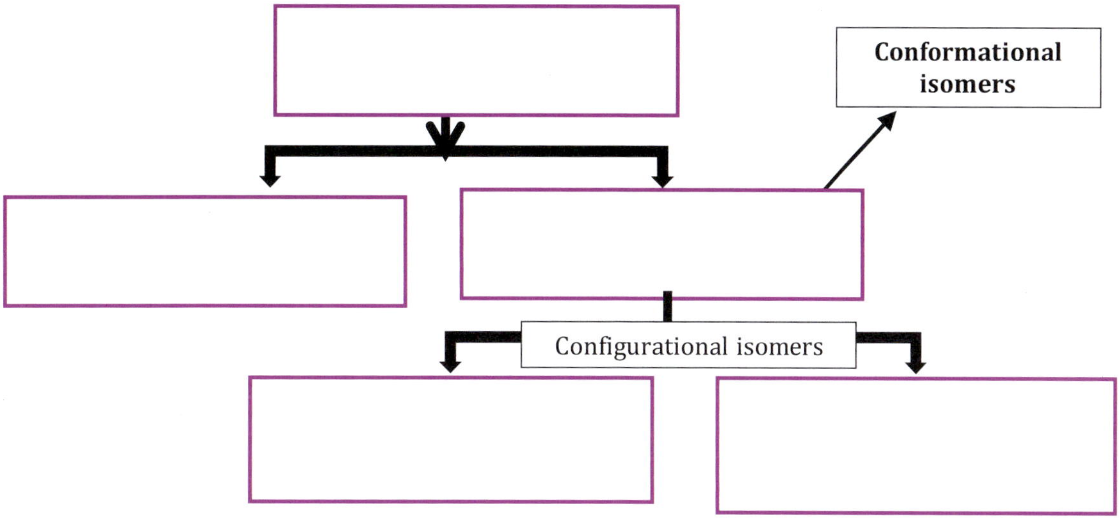

Notes

Lecture Topic I.19-22: Stereochemistry
Definitions

Stereoisomers:

(A)

Conformational stereoisomers:

(B)

- staggered and eclipsed conformers
- *gauche* and *anti* conformers
- Chair or boat chair conformers of cyclohexane

Configurational stereoisomers:

(C)

NOW:

enantiomers:

(D)

diastereomers:

(E)

Notes

Lecture Topic I.19-22: Stereochemistry
Chirality

Molecules with "handedness" possess (A) [ ]

and are (B) [ ]

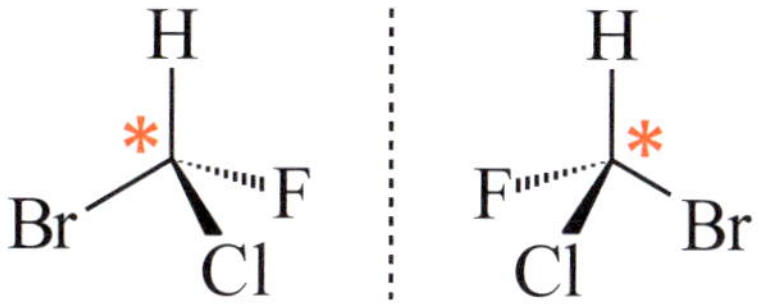

These are nonsuperimposable mirror images; using definitions on the previous slide, this is a pair of **enantiomers** ("left- and right-handed" isomers). The molecule is chiral and possesses a **chiral** (**stereogenic**) **center** (also called a **stereocenter**). These centers can be labeled with asterisks for clarity.

(C) [ ]

Notes

Lecture Topic I.19-22: Stereochemistry
Identifying Stereocenters and Chirality

(A) For a carbon atom to be a stereocenter it must:

(B)

A molecule WITHOUT a stereocenter is

(C)

If the molecule has a **plane of symmetry** it is

(D)

A molecule with a stereocenter **and** a plane of symmetry is achiral.

(E) example:

Such compounds are called **meso** compounds.

Notes

Lecture Topic I.19-22: Stereochemistry
Identifying Stereocenters

Mark the stereocenters with asterisks.

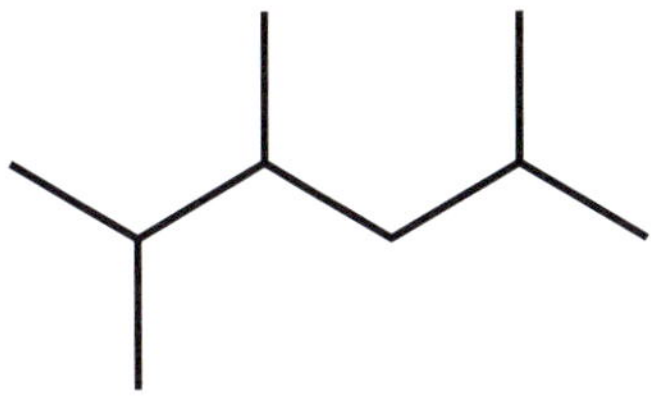

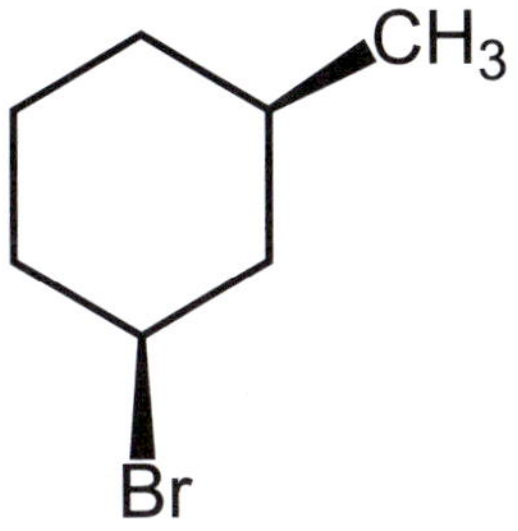

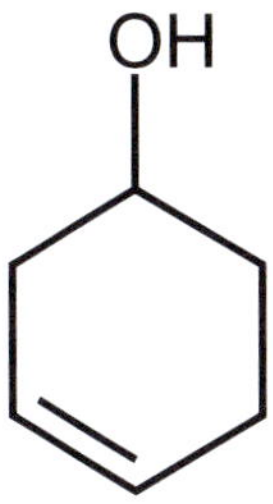

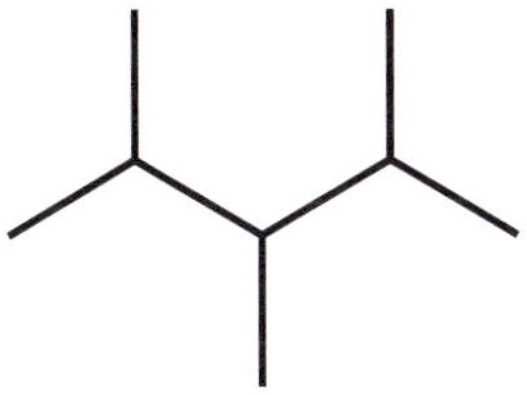

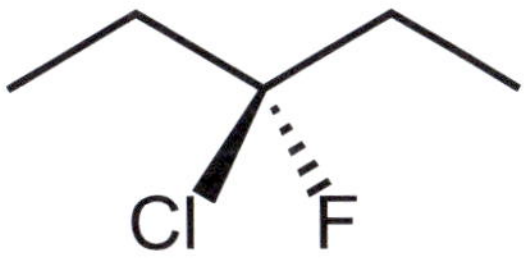

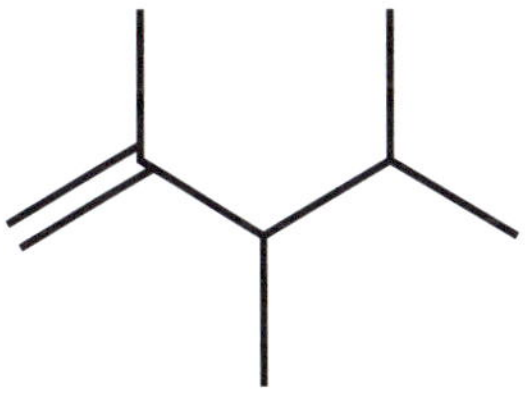

Notes

Lecture Topic I.19-22: Stereochemistry
Properties of Enantiomers

Many physical properties of enantiomers are **IDENTICAL**

(A)

Two Enantiomers can interact very differently with other chiral molecules. **This is especially important in biological systems!**

1. Amino acids

(B)

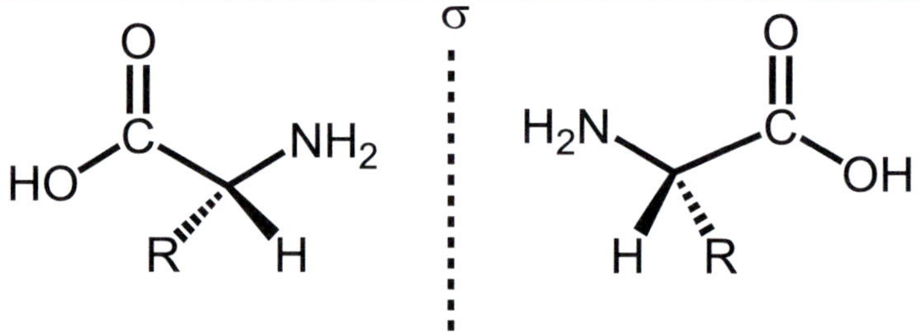

Notes

Lecture Topic I.19-22: Stereochemistry
Properties of Enantiomers

2. Enzymes/cellular signaling

(A)

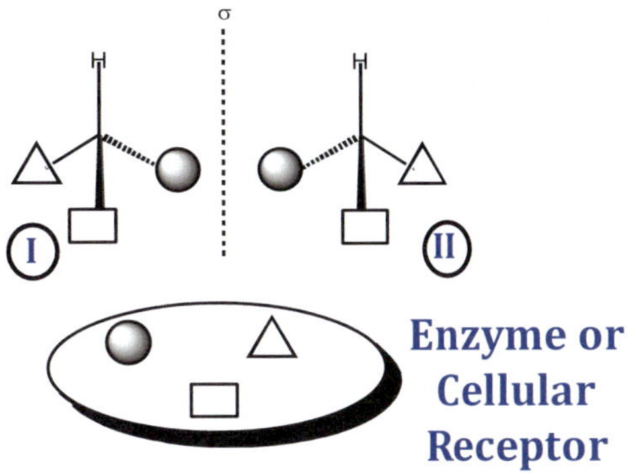

3. Drugs

(B)

Notes

Lecture Topic I.19-22: Stereochemistry

Distinguishing Enantiomers

If many physical properties are identical, we use to distinguish two enantiomers?

(A)

(B)

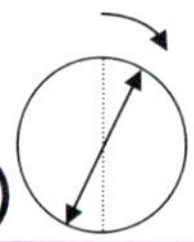

(C)

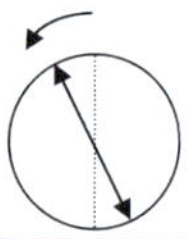

(D)

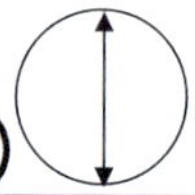

Notes

Lecture Topic I.19-22: Stereochemistry
Vocabulary

Optically inactive:

(A)

Optically pure: (C)

Optical purity: (B)

(a sample with 80% enantiomer '**X**' and 20% enantiomer '**Y**' has 60% e.e. or 60% optical purity)

Notes

Lecture Topic I.19-22: Stereochemistry
Types of Isomerism

Enantiomers of different compounds rotate light by different amounts, so an optically pure compound has a characteristic **specific rotation [α]**. Two enantiomers of the same compound have **specific rotations** with the same magnitude but in opposite directions. The equation used to calculate the specific rotation is:

$$[\alpha]_D^T = \frac{\alpha}{cl}$$

- temperature (T)
- observed rotation (degrees) (α)
- path length (dm) (l)
- concentration (g/mL solution) (c)
- Type of light used (sodium D line) (D)

(+)-2-butanol $[\alpha]_D^{27}$ = + 13.5°
(–)-2-butanol – 13.5°

Notes

Lecture Topic I.19-22: Stereochemistry
Absolute Configuration is Expressed with R- *or* S-

We obviously need a convention to designate the **absolute configuration** of a given molecule.

A

B

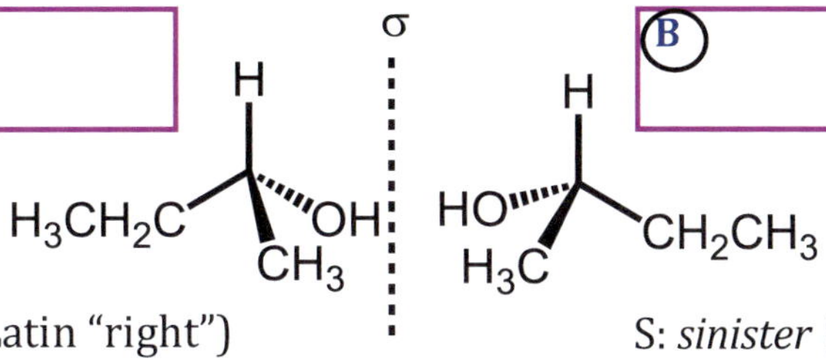

R: *rectus* (Latin "right") S: *sinister* (Latin "left")

Whether the configuration of a chiral center is *R* or *S* depends only on

C

and is **NOT** determined by

D

So, we need a convention for assigning "priority" to the units that are attached to a chiral atom ...

Notes

Lecture Topic I.19-22: Stereochemistry
Cahn-Ingold-Prelog (CIP) Convention

Priority Rules:

1. First look at atom directly attached to the stereogenic atom

(A)

2. If same atomic number, higher mass = higher priority (D > H, $^{13}C > ^{12}C$, etc.)

3. If atoms A and B are identical, move to highest priority atom attached to A and B until a break in the tie is found (first compare A1 to B1 priority. If tie, compare A2 to B2. If tie, compare A3 to B3, etc.)

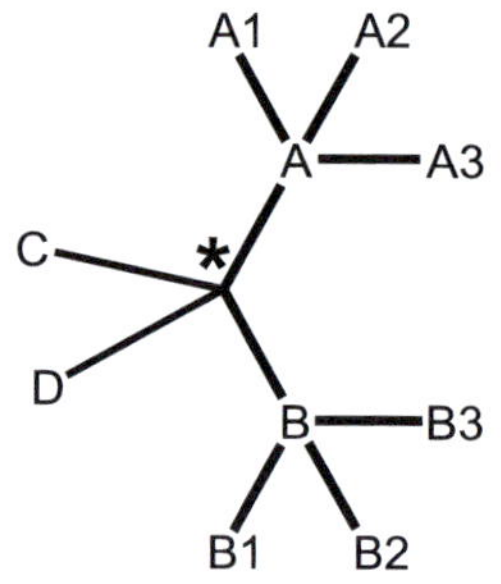

4. For Example:

(B)

Notes

Lecture Topic I.19-22: Stereochemistry
Cahn-Ingold-Prelog (CIP) Convention

4. If a substituent is doubly or triply bonded to another atom, use the 'break and duplicate' strategy to create 'false atoms' as a visual aid to prioritize:

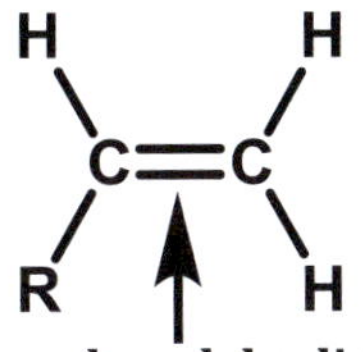

break and duplicate

Becomes

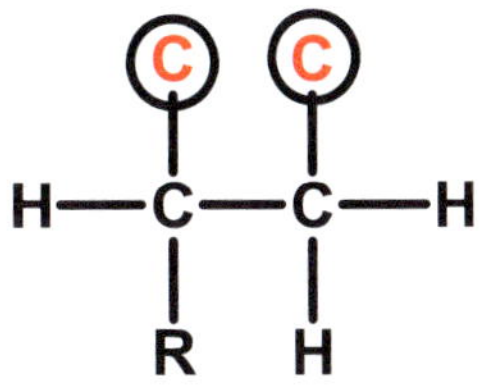

A

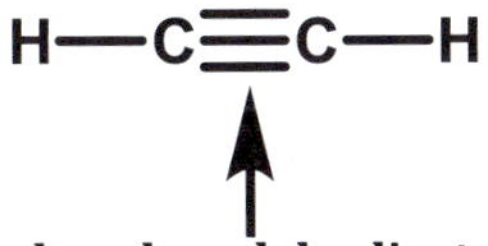

break and duplicate

Becomes

B

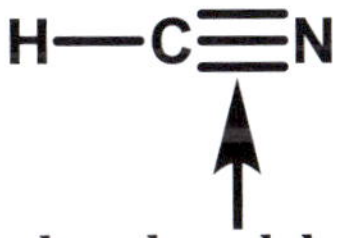

break and duplicate

Becomes

Notes

Lecture Topic I.19-22: Stereochemistry
Newman Projections Conveniently Show Conformation

1. (A)
   (using CIP rules)
2. Point the **lowest** priority group away from you

3. Determine the direction of procession from **1-2-3** priorities:

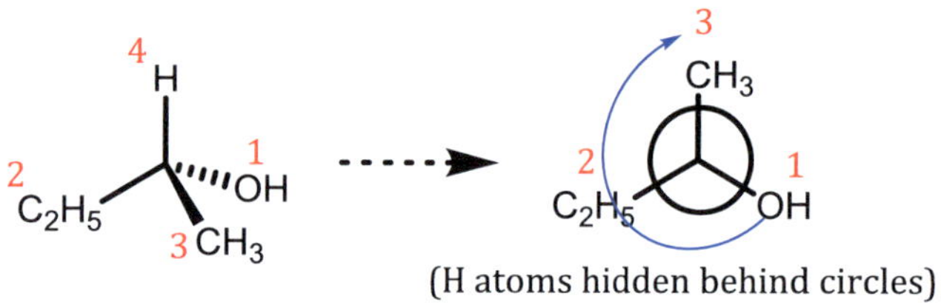

Clockwise 1-2-3:

(B)

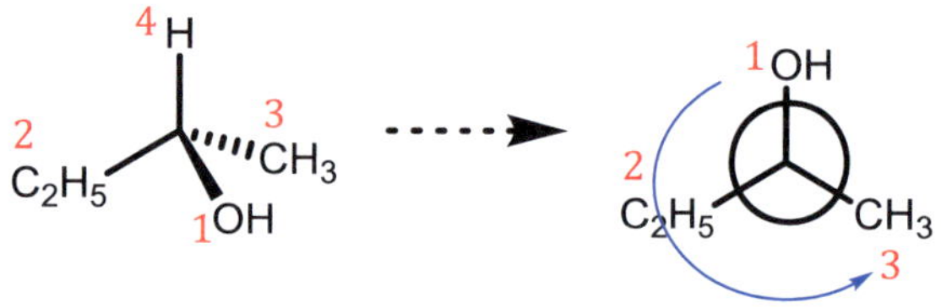

Counterclockwise 1-2-3:

(C)

Notes

Lecture Topic I.19-22: Stereochemistry

There are Many Ways to Draw a Chiral Object

Introducing the Creepy Chiral Stickman!

<u>Chiral Athlete Doing Pushups</u>

Sweet tattoo on right hand

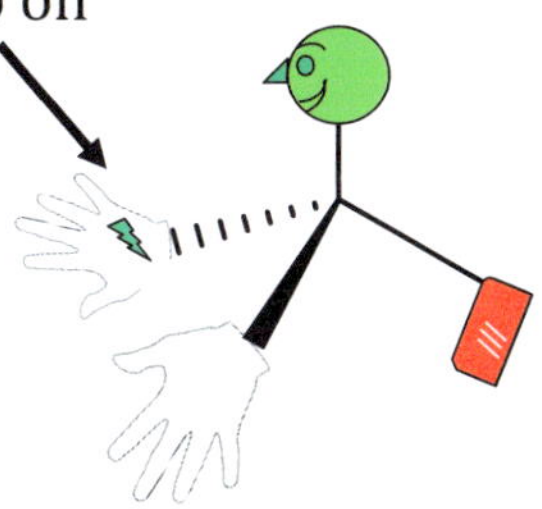

<u>Chiral Molecule … Doing Pushups (?)</u>

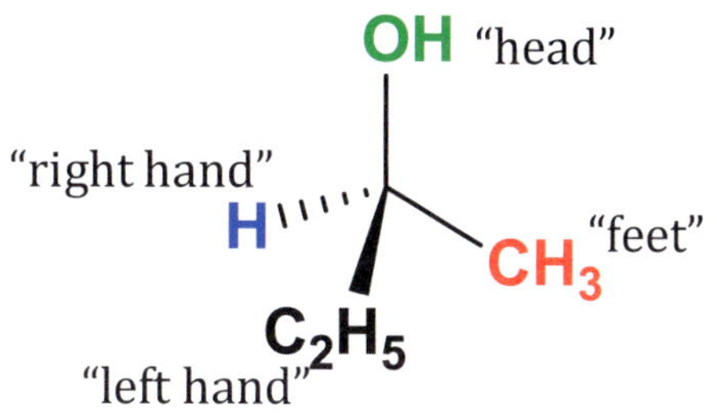

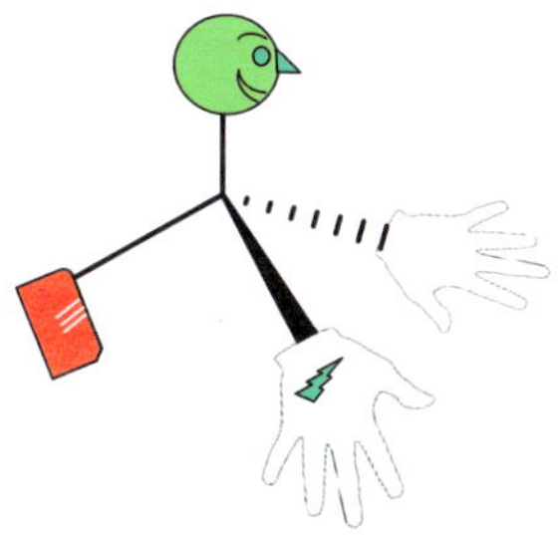

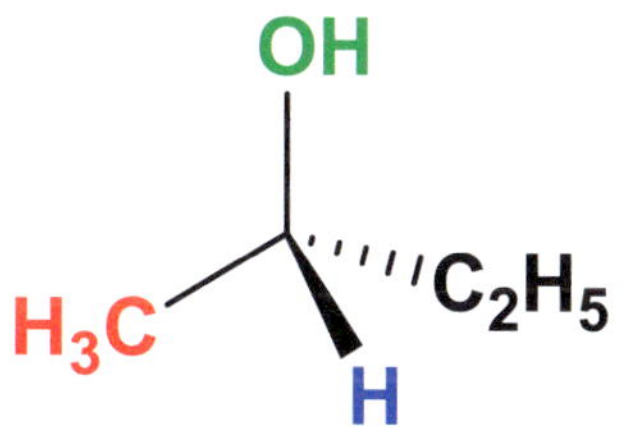

We need a way to visualize rotation of a chiral molecule without losing stereochemical information

<u>Notes</u>

Lecture Topic I.19-22: Stereochemistry
There are Many Ways to Draw a Chiral Object

Hand-Stand

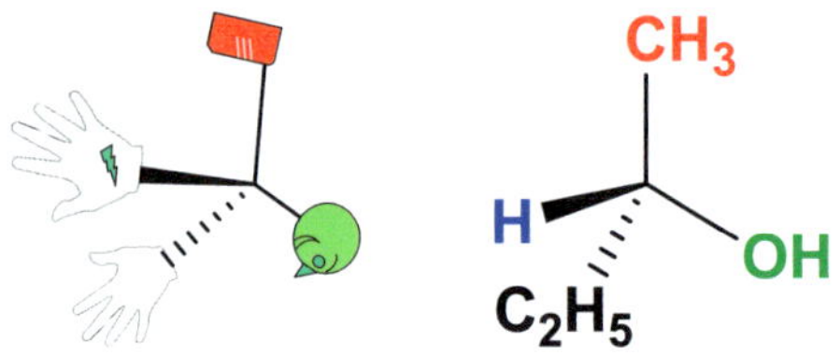

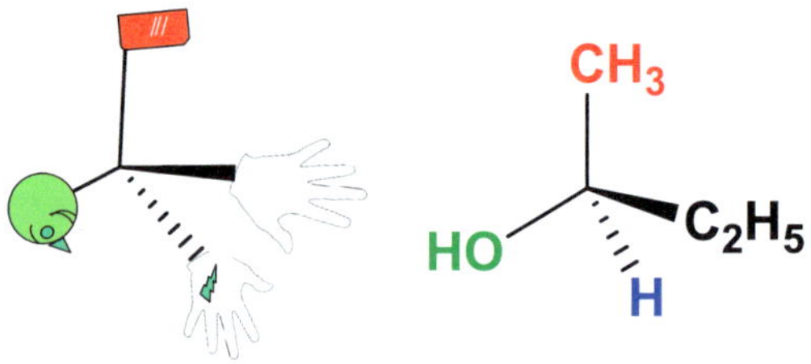

Ceiling Pushup!

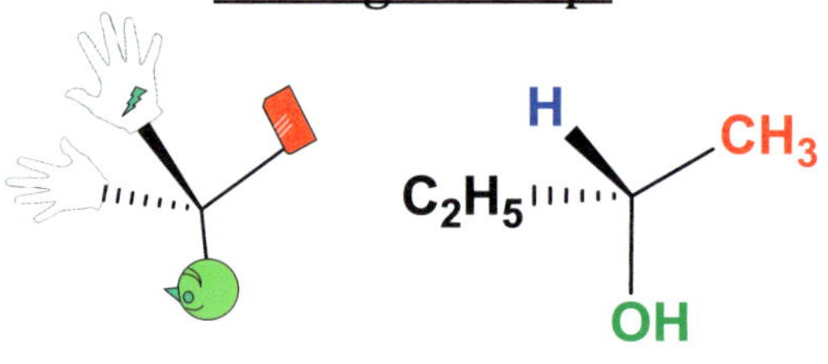

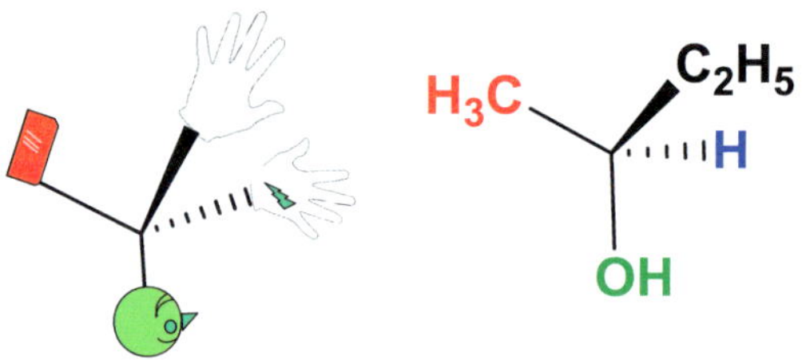

Notes

Lecture Topic I.19-22: Stereochemistry

Many Ways to Draw (S)-2-butanol

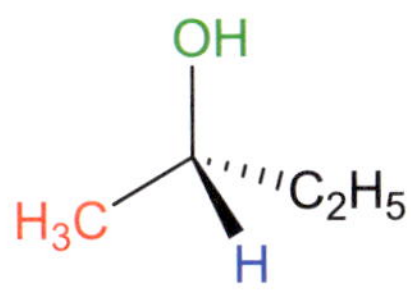

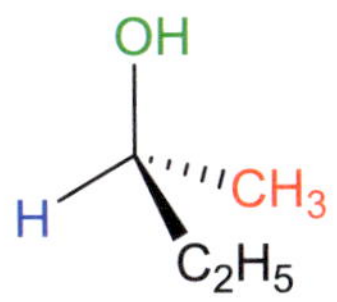

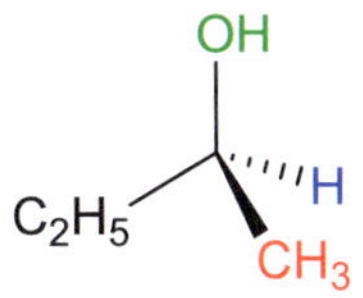

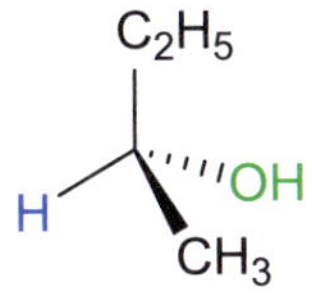

A Two other ways:

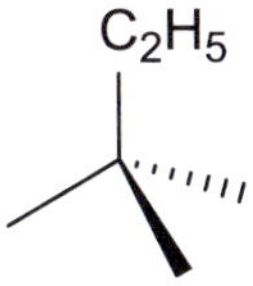

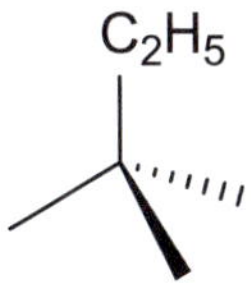

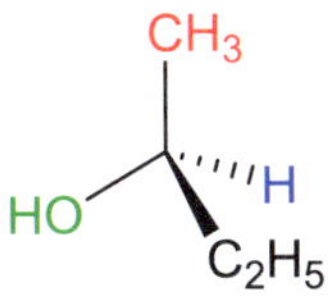

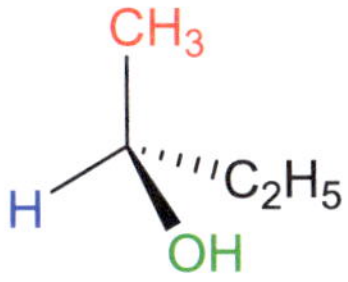

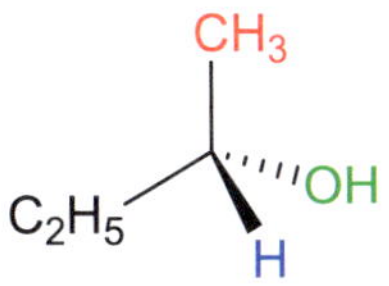

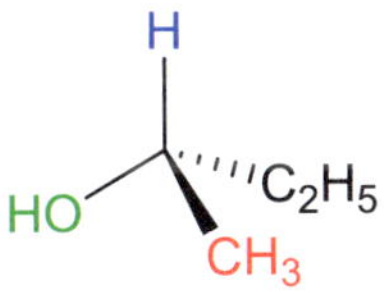

B Two other ways:

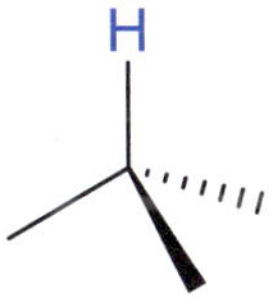

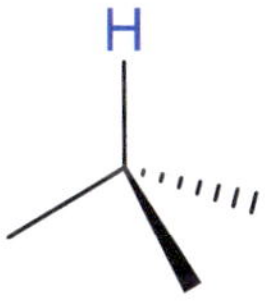

Notes

Lecture Topic I.19-22: Stereochemistry
Fischer Projections

In addition to standard dash-wedge notation, a **Fischer projection** can show 3D shape.

In a Fischer projection, horizontal lines: (A)

Vertical lines: (B)

Fischer Projection

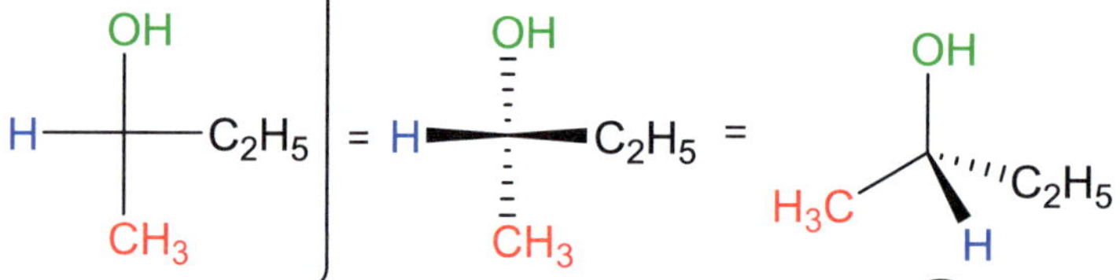

(One of many ways to draw it)

Coming in for a hug but leaning away
(tattoo on palm as well apparently)

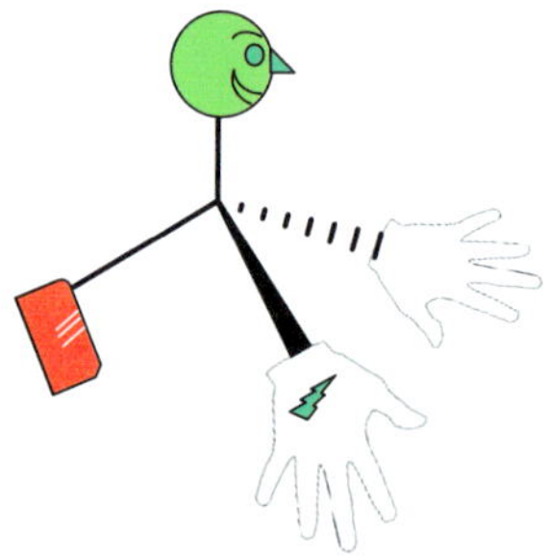

Notes

Lecture Topic I.19-22: Stereochemistry

Determining R and S with Fischer Projections

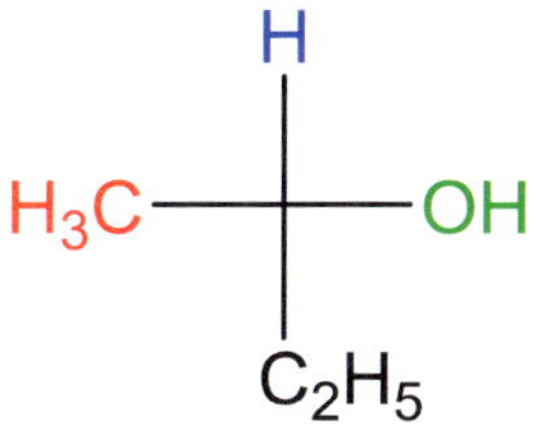

(A) Name:

Convert to wedge and dash structure (several correct representations):

(B)

Fischer Projection and Wedge/Hashed Lines for (*R*)-3-methylheptane:

(C)

Notes

Lecture Topic I.19-22: Stereochemistry

Multiple Stereocenters: Diastereomers

Diastereomers:

(A)

A compound with 'n' chiral centers can have up to 2n stereoisomers.

Consider 3-chloro-2-butanol, which has Stereocenters at C2 and C3:

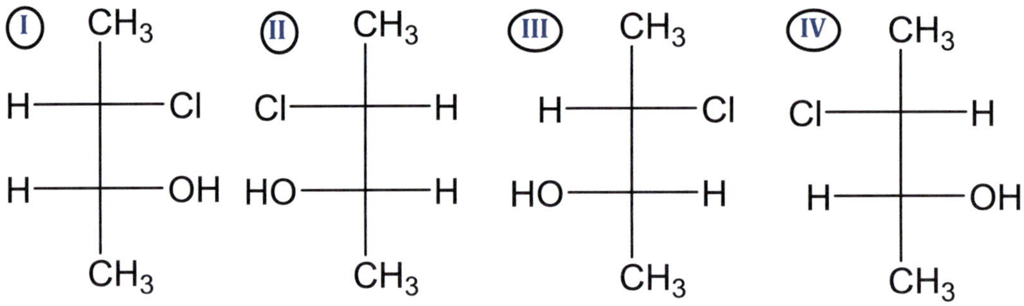

I and II = (B)

III and IV = (C)

Any other pair = (D)

Notes

Lecture Topic I.19-22: Stereochemistry
Meso Compounds

Meso Compounds:

(A)

So compounds I and II are ACHIRAL!

Consider 2,3-dichlorobutanol, which has Stereocenters at C2 and C3:

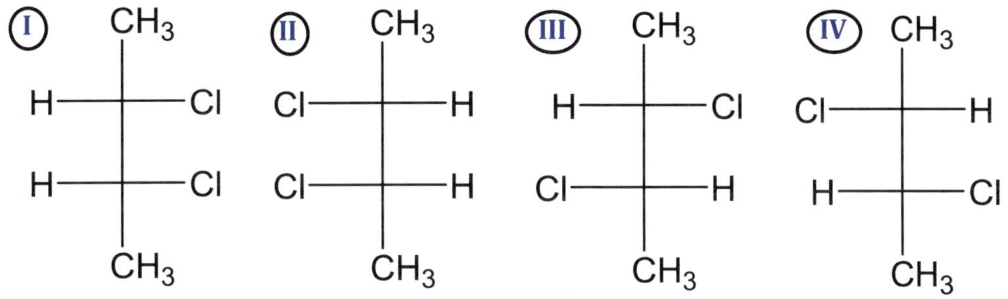

I and II = (B)

Any other pair =

III and IV = (C)

(D)

Notes

PART II: Substitution, Elimination and Oxidation

Lecture Topics in Part II:

1. Stereochemistry V: Tracking Stereocenters in Reactions
2. Nucleophilicity versus Basicity
3. Substitution Reactions of RX I: The S_N1 Mechanism
4. Substitution Reactions of RX II: The S_N2 Mechanism
5. Substitution Reactions of RX III: Stereochemistry of Substitution Reactions
6. Elimination Reactions of RX I: The E1 Mechanism
7. Elimination Reactions of RX II: The E2 Mechanism
8. Stereoelectronic Effects in the E2 Reaction
9. Factors Leading to Non-Zaitsev Products in the E2 Reaction: Hofmann Elimination
10. Competition Among Substitution and Elimination Reactions of RX: Predicting Pathways
11. Reactions of Alcohols I: Substitution and Elimination
12. Oxidation and Reduction: Definitions
13. Reactions of Alcohols II: Oxidation
14. Acid Cleavage of Ethers
15. Ring-Opening of Epoxides: Steric versus Electronic Effects on Product Distribution

Lecture Topic II.1: Tracking Stereocenters in Reactions

Tracking Stereocenters in Reactions

Stereocenters may be generated, eliminated or changed in the course of a given reaction on the basis of the mechanism involved in the reaction.

A couple of overarching concepts can help keep track of Stereocenters in a reactions:

1. If reagents/reactants are achiral, then:

(A)

2. If reagents/reactants are chiral, then:

(B)

Notes

Lecture Topic II.1: Tracking Stereocenters in Reactions
Tracking Stereocenters in Reactions

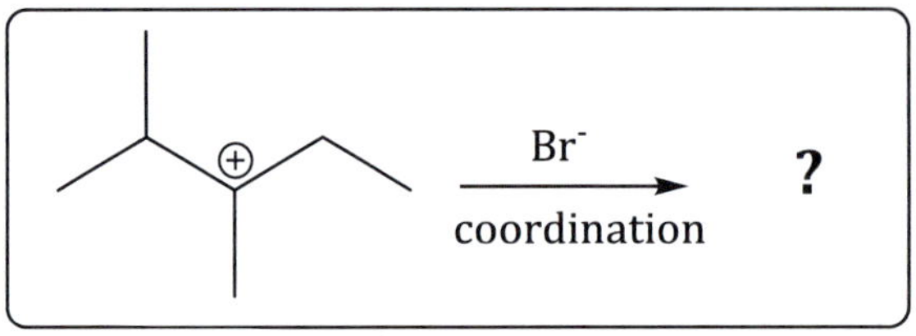

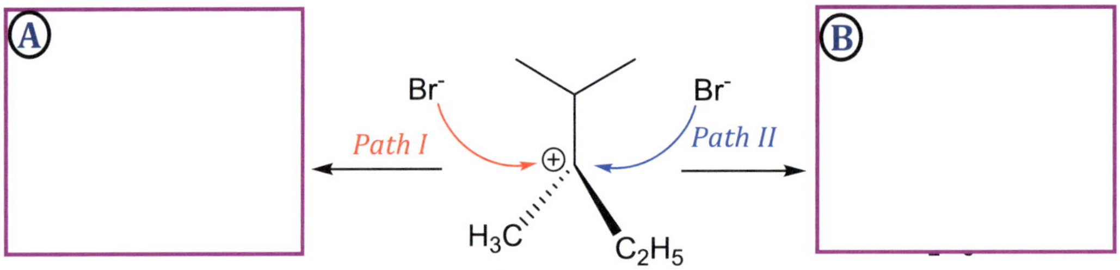

R-isomer *Achiral Reactants!* *R*-isomer

The steric barrier to coordination from either face is the same. The two enantiomers have equal stability. So the product is a:

Ⓒ

We will apply our knowledge of these general principles to other reactions as we encounter them.

Notes

Lecture Topic II.2: Nucleophilicity versus Basicity

Thermodynamics versus Kinetics

Nucleophilicity is a measure of how rapidly a compound (a nucleophile) will attack an electron-deficient atom

Nucleophilicity is measured by a rate constant (*k*).

(A)

Basicity is a measure of how favorable it is for a compound (a base) to shares its lone pair, often with a proton.

Basicity is measured by the acid dissociation constant (K_a).

(B)

Notes

Lecture Topic II.2: Nucleophilicity versus Basicity
Nucleophilicity Trends

Nucleophilicity Trends

Ⓐ

$O^{2-} > HO^- > H_2O$

$RO^- > ROH$

$HS^- > H_2S$

Ⓑ

$H_2N^- > HO^- > F^-$

$H_3N > H_2O$

$RO^- > RCO_2^-$

Ⓒ

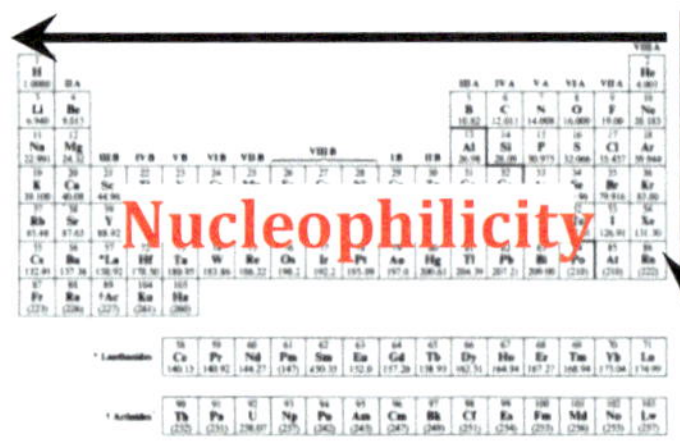

$I^- > Br^- > Cl^- > F^-$

$HS^- > HO^-$

$PH_3 > NH_3$

Notes

Lecture Topic II.2: Nucleophilicity versus Basicity

Solvent Influences Nucleophilicity/Basicity

(A)

(B)

It is easier to break the ion-dipole interactions between a weak base and the solvent than between a strong base and the solvent

(C)

Notes

Lecture Topic II.2: Nucleophilicity versus Basicity

Classifying Solvents

Solvents can be classified into several types on the basis of the type of intermolecular forces they can exhibit with other molecules:

(A) hexane, other hydrocarbons

(B) diethylether, acetone

polar protic: (C)

(D) $(CH_3)_2SO$ (dimethylsulfoxide, DMSO)
$(CH_3)_2NC(O)H$ (N,N-dimethylformamide, DMF)

Notes

Lecture Topic II.2: Nucleophilicity versus Basicity
Solvent Influences Nucleophilicity/Basicity

A

OH^- in H_2O: strong ion-dipole interaction with water diminishes its nucleophilicity (less reactive; higher E_a)

in DMSO: weaker solvation and thus more nucleophilic (faster reaction) versus the same reaction in water

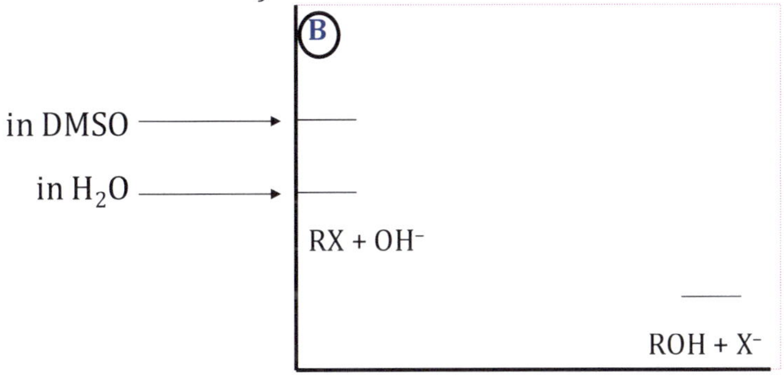

Notes

Lecture Topic II.3: The S_N1 Mechanism

Alkyl Halides can undergo the S_N1 Reaction

Alkyl Halides react with even poor nucleophiles, often solvents such as water or alcohols, to yield alcohols or ethers:

$$\text{H-OH} + \text{R-X} \longrightarrow \text{R-OH} + H^+ + X^-$$

$$\text{H-OR'} + \text{R-X} \longrightarrow \text{R-OR'} + H^+ + X^-$$

This is a **substitution reaction**, and it proceeds by a specific mechanism called the **S_N1** mechanism. This is an abbreviation meant to help you remember how the reaction takes place:

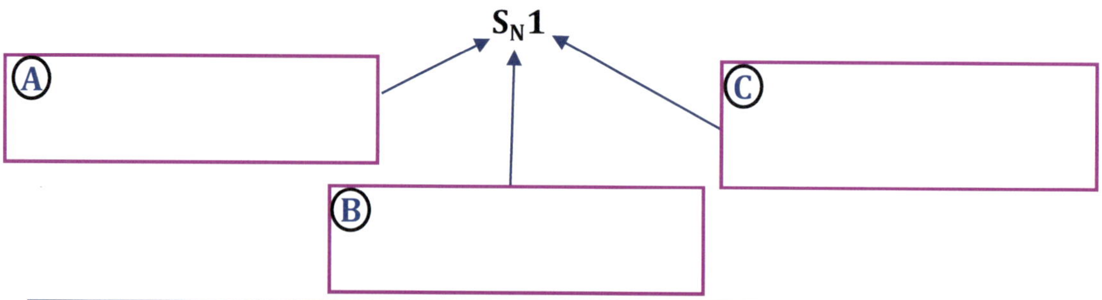

Because the substrate (RX) reacts with the solvent, this reaction is sometimes called **solvolysis** as well.

Notes

Lecture Topic II.3: The S_N1 Mechanism

S_N1 is Heterolysis then Coordination

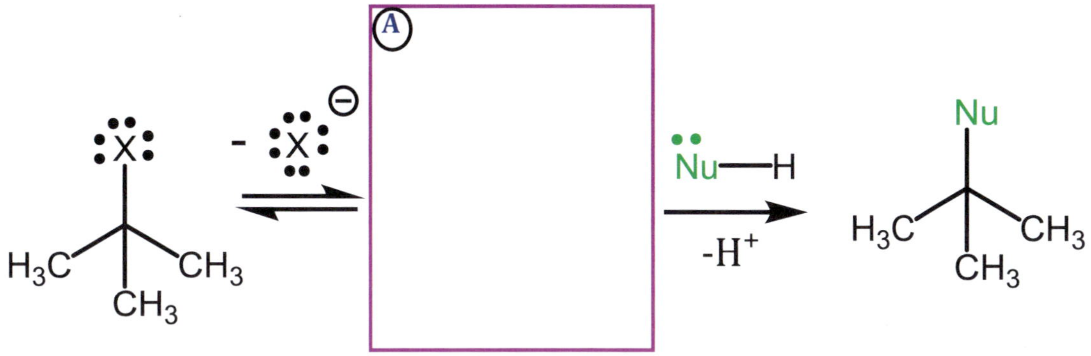

S_N1 is a two-step sequence of steps:

1) Heterolysis (lose a good **Leaving Group (LG)**)
2) Coordination (of a Nucleophile)

Heterolysis occurs: (B)

A "good" leaving group (LG) is: (C)

Notes

Lecture Topic II.3: The S_N1 Mechanism

S_N1 is Heterolysis then Coordination

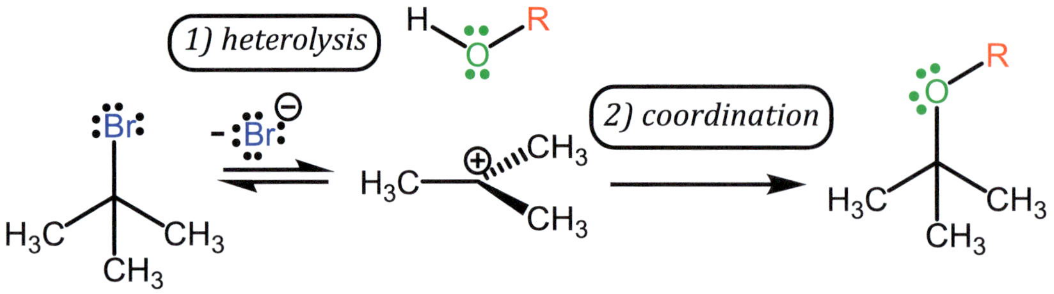

S_N1 = 1) Heterolysis, 2) Coordination. **Fill in the curved arrows to show the flow of electrons**

R in HOR can be: (A)

Rate-Limiting Step: (B)

Rate Law: (C)

Notes

Lecture Topic II.3: The S_N1 Mechanism
Factors Influencing S_N1 Rate

Rate-Limiting Step for S_N1

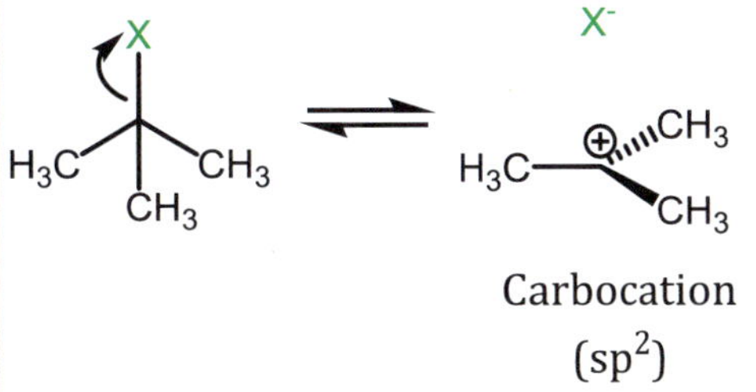

The rate-limiting step leads to a leaving group (here X-) and a carbocation as products. Substrates that produce more stable products react faster by S_N1.

Substrate trend: $CH_3X < 1^\circ$ RX < [ ] < [ ]

Too unstable; will not work for S_N1!

REACTIVITY →

Leaving group (LG) trend: R-F < [ ] < [ ] < [ ]

Generally difficult for S_N1!

Notes

Lecture Topic II.3: The S_N1 Mechanism

Factors Influencing S_N1 Rate

(A)

Solvent can stabilize cations

(B)

Solvent can stabilize anions
(we saw this for bases)

Stronger attractive intermolecular forces are more stabilizing, so the solvents in which the S_N1 reaction is the fastest are:

(C)

Notes

Lecture Topic II.3: The S_N1 Mechanism
Carbocation Rearrangement to Gain Stability

Thermodynamically favorable reactions will occur spontaneously when they are kinetically and mechanistically accessible. In **Lecture Topic I.8** we saw that carbocations can rearrange.

Carbocations will always

(A)

The rate at which the carbocation rearranges is:

(B)

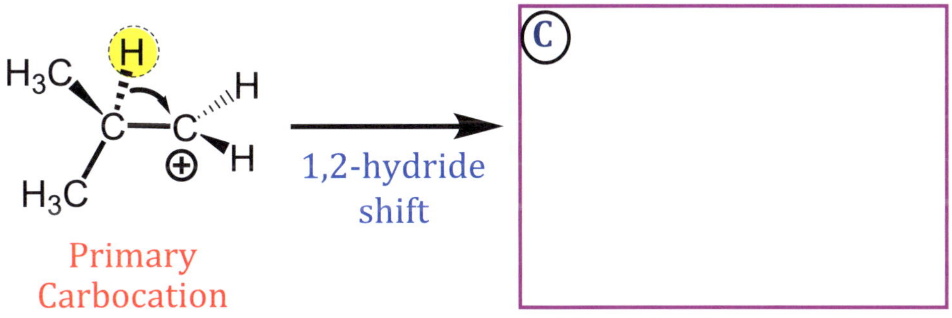

Notes

Lecture Topic II.3: The S_N1 Mechanism
Carbocation Rearrangement to Gain Stability

Alkyl groups can also migrate to create a more stable species. This *usually* only happens when there are no hydride units to shift. A notable exception is when rearrangement leads to relief of ring strain (given as an Example in the Reaction Guide).

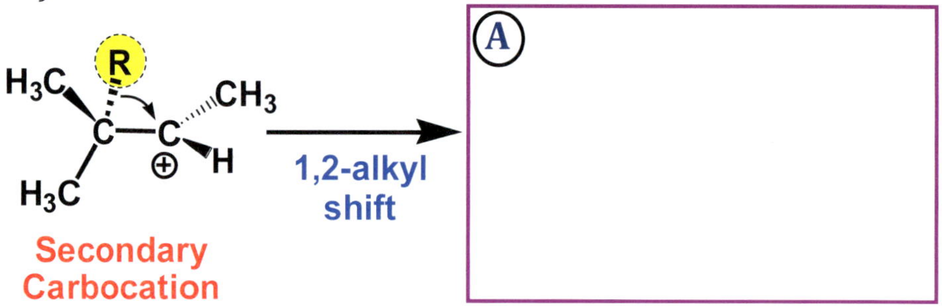

Because a carbocation forms as an intermediate in the S_N1 reaction:

B

Notes

Lecture Topic II.3: The S_N1 Mechanism

The Major S_N1 Product derives from the Most Stable Carbocation

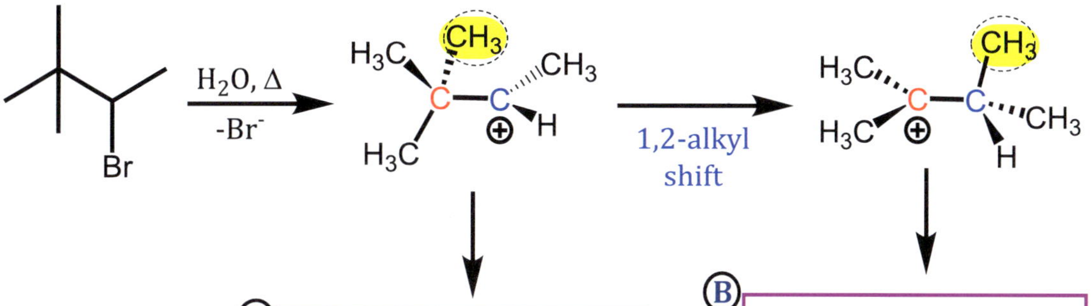

(A)

3% of substitution

(B)

97% of substitution

(C) The Major Product:

Notes

Lecture Topic II.4: The S_N2 Mechanism

S_N2 is a Concerted, Bimolecular Reaction

We learned the general form of the S_N2 reaction in **Lecture Topic 1.8**. We will now study this mechanistic step is more detail.

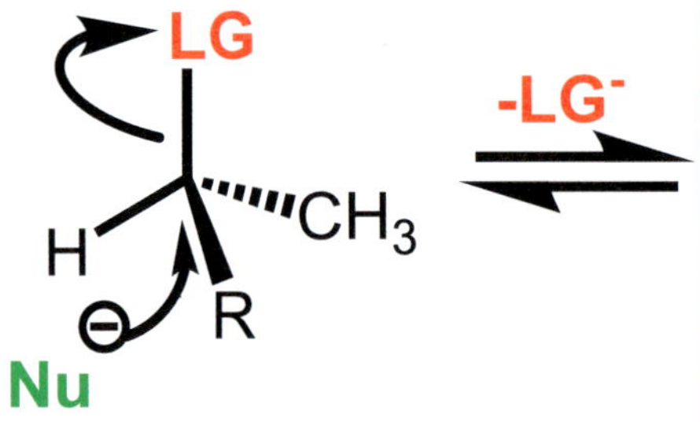

A

B Unlike the S_N1 reaction, the S_N2 reaction occurs in a single step. This type of reaction is called a:

C That there is only one step means that Nu, substrate and leaving group all influence rate. The rate law is:

Like S_N1, then, more stable LG = faster reaction rate

Notes

Lecture Topic II.4: The S_N2Mechanism

S_N2 Requires a "Good" Nucleophile

In S_N1, the nucleophile attacks a cation, which has a strong enough pull on electrons to pull in even poor nucleophiles. In S_N2, the nucleophile attacks a neutral compound, which has a significantly lower attraction for a nucleophile than does a cation. For this reason:

(A)

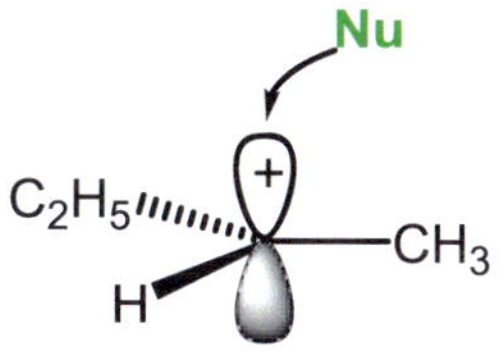

Nucleophilic attack in the coordination step of the $S_N1$mechanism

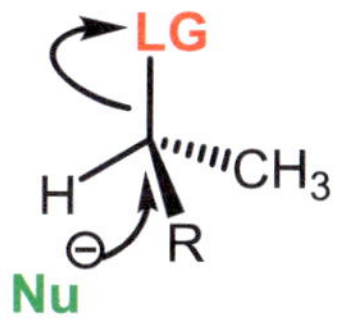

Nucleophilic attack in the concerted $S_N2$mechanism

(B) The better the Nucleophile:

Notes

Lecture Topic II.4: The S_N2Mechanism

Good Nucleophiles for S_N2

What structural features do we look for in order to identify 'good' nucleophiles for the S_N2 reaction?

(A)

General form of "Bulky" Poor Nu

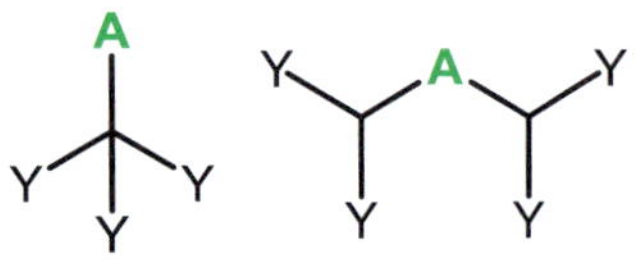

three branches *four branches*

A = Nucleophilic site, Y = non-H branch

Examples of Poor Nu

F^-, H_2O, HOR

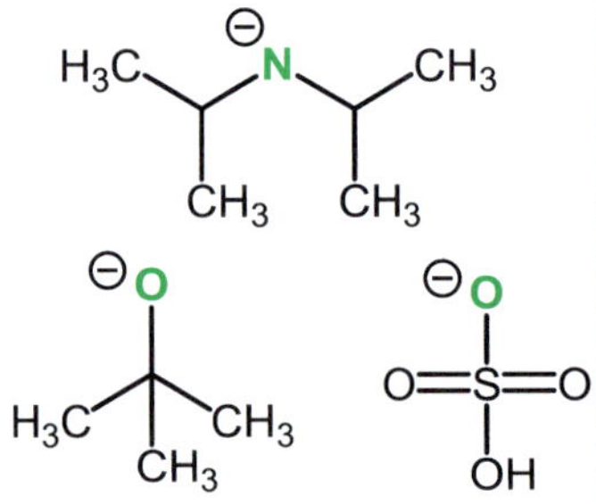

Examples of Good Nu

Cl^-, Br^-, I^-, HO^-

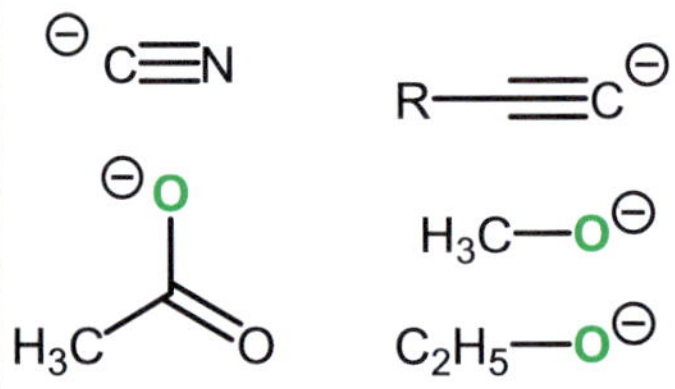

RSH or PR_3

(R = primary, secondary)

Notes

Lecture Topic II.4: The S$_N$2Mechanism

Bulky Substrates Slow S$_N$2 Rate

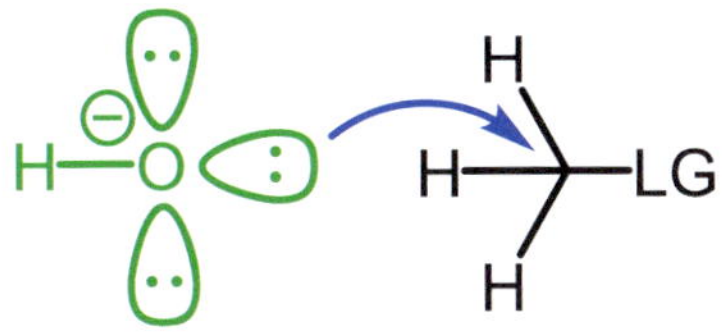

Ⓐ

Ⓑ

Similar to how steric hindrance slows bulky Nucleophiles

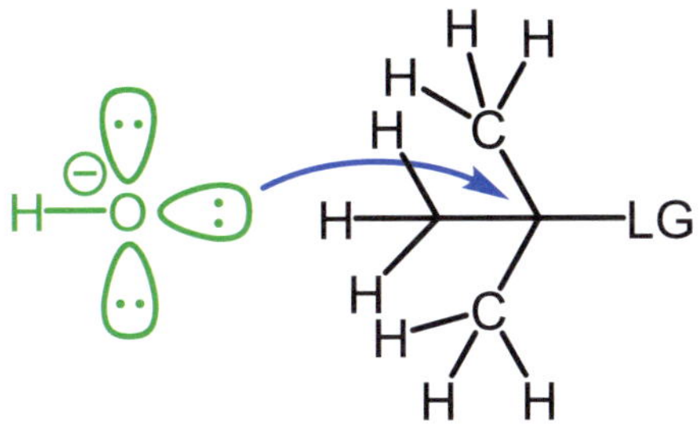

Ⓒ This observation about sterics allows us to predict S$_N$2 rates:

Tertiary halides do not work for S$_N$2 due to sterics!

Notes

Lecture Topic II.4: The S_N2 Mechanism

Solvent Influences S_N2 Rate

(A)

Solvent weakens Nu strength

(B)

Solvent can stabilize anions
(we saw this for bases)

The influence on Nu strength influences the rate to a greater extent than does the stabilization of the anion, so:

(C)

Notes

Lecture Topic II.5: Stereochemistry of Substitution Reactions

Stereochemistry of S_N1 Reactions

Recall from **Lecture Topic II.1** that a single chiral product cannot be obtained from achiral starting materials for **any** step of a reaction sequence. Consider the reaction of a chiral alkyl halide by an S_N1 pathway:

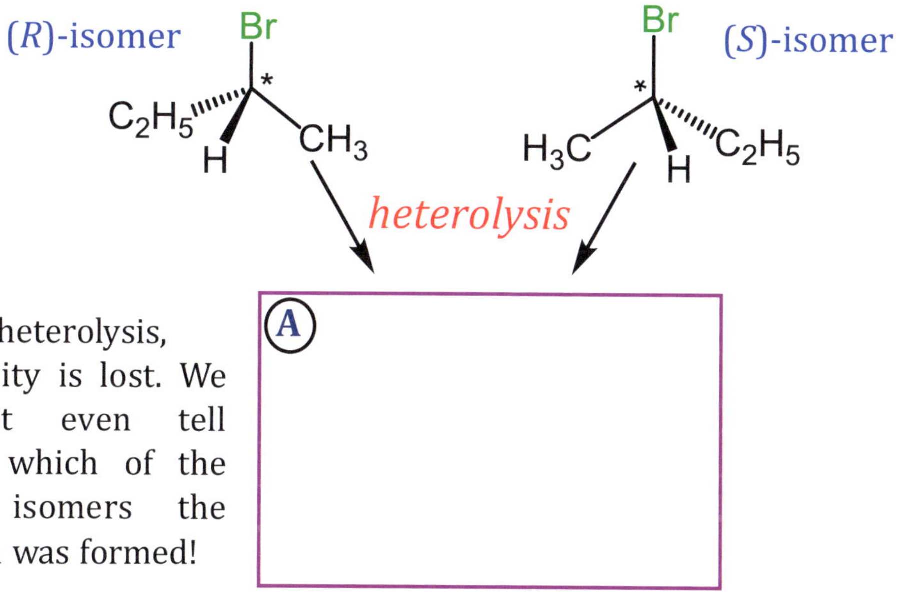

After heterolysis, Chirality is lost. We cannot even tell from which of the two isomers the cation was formed!

An Achiral Carbocation

Notes

The carbocation formed by heterolysis has a trigonal planar geometry; it is achiral and symmetric. The nucleophile may attack from either face of the plane with equal probability:

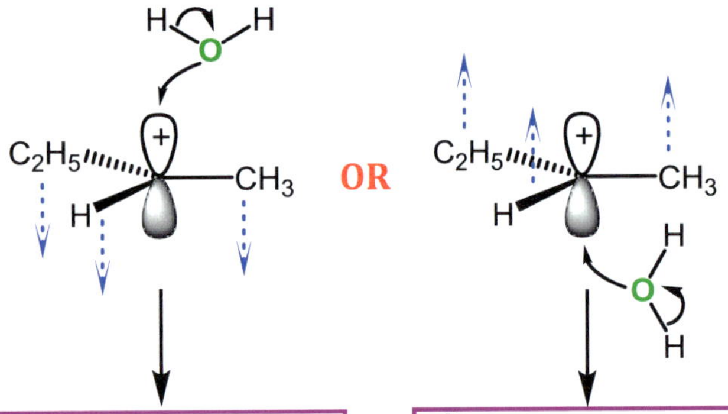

(A) 50%

(B) 50%

(C) The product is a:

(D) The S_N1 rxn is:

Notes

The S_N2 reaction is concerted, so if the nucleophile substitutes at a stereogenic site, a single chiral compound will be produced:

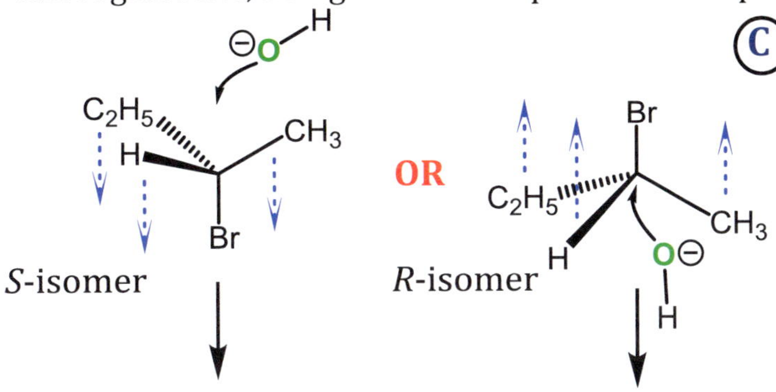

(A)

(B)

(C) The product is:

A Process called:

(D) The S_N2 rxn is:

Notes

Lecture Topic II.6: The E1 (Unimolecular Elimination) Mechanism
E1 is Heterolysis then Electrophilic Elimination

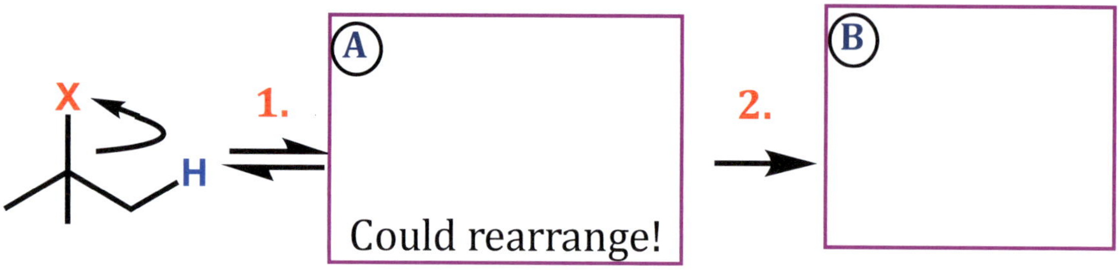

This is the E1 reaction: Heterolysis then electrophilic elimination. **Heterolysis is rate-limiting**, as in S_N1. The intermediate carbocation can also rearrange. In fact, many aspects of E1 are similar to S_N1:

Rate Law: (C)

Substrate: (D)

Leaving Group: (E)

Solvent: (F)

Notes

Lecture Topic II.6: The E1 Mechanism
Determining Major E1 Product

Once heterolysis occurs to generate the carbocation intermediate, there may be more than one possible electrophilic elimination to consider:

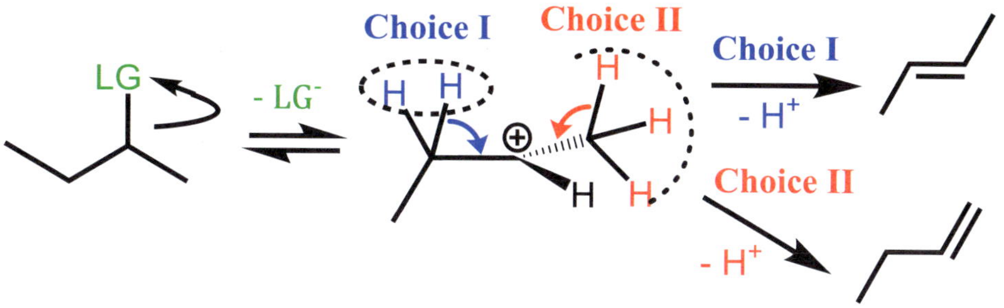

We know that stability is important in predicting product distribution. Alkene stabilities have been measured (more on how in **Lecture Topic III.2**) and found follow this trend (R = a hydrocarbon substituent):

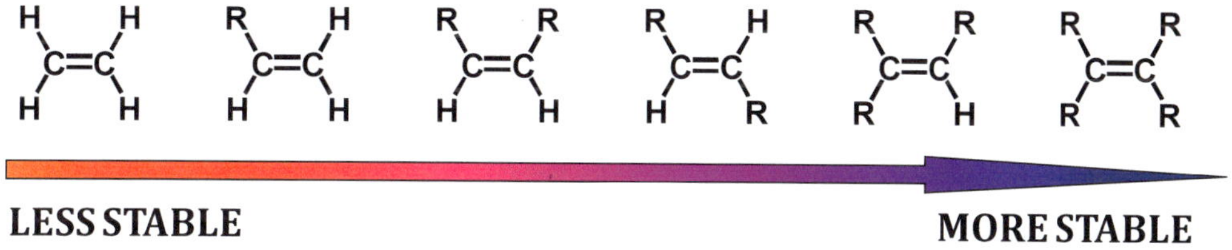

Notes

Lecture Topic II.6: The E1 Mechanism

Regioselectivity and Zaitsev's Rule: Reaction Coordinate Diagram

Observation = "Zaitsev's Rule:

Ⓐ

Explanation = Reaction coordinate diagram:

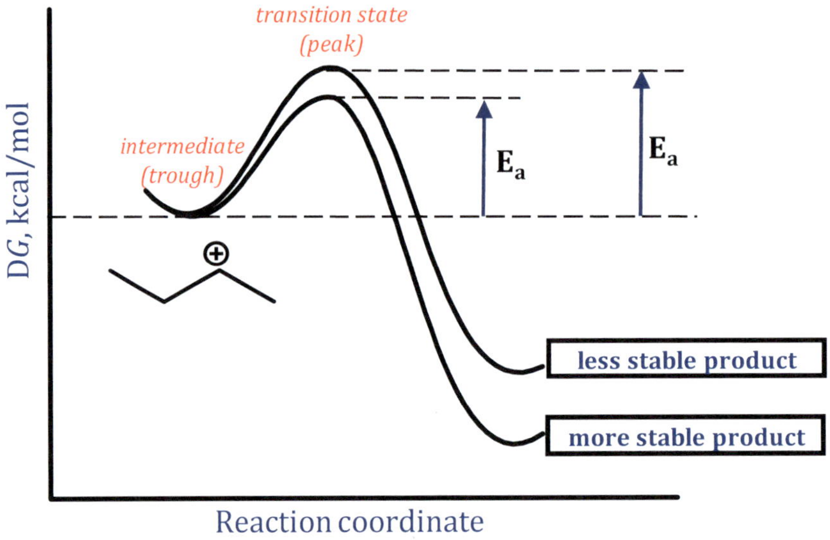

Notes

Lecture Topic II.6: The E1 Mechanism

Conformational Analysis Helps Determine Product Configuration

A

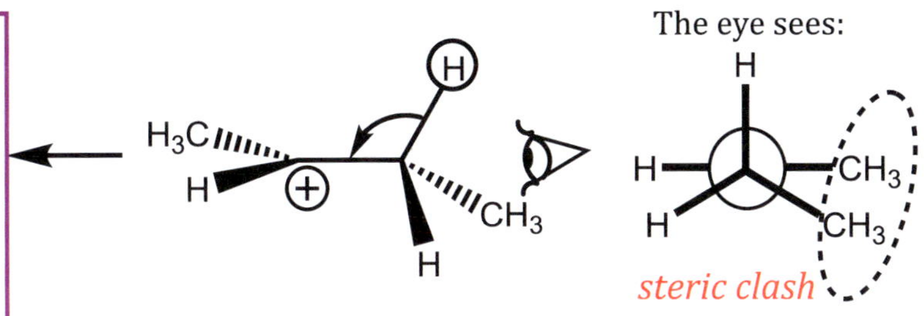

B

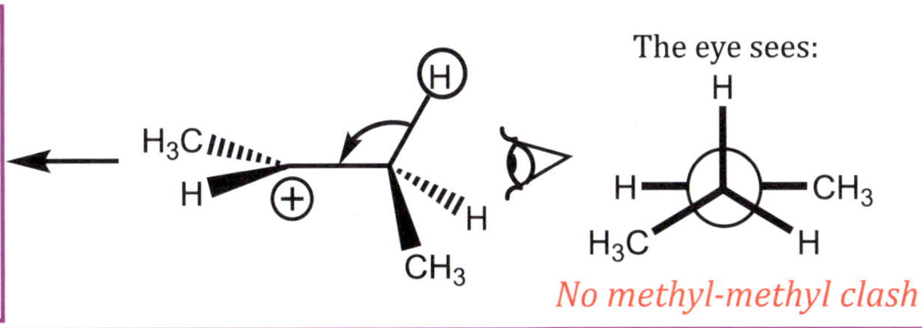

C

Notes

Lecture Topic II.6: The E1 Mechanism
Zaitsev's Rule in Action

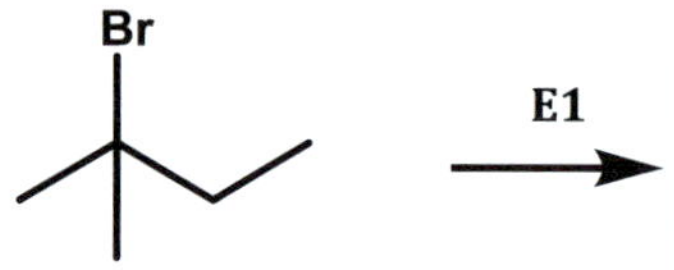

A	+	B
90%		10%
More Stable		**Less Stable**

C	D	E
56%	32%	12%
MostStable		**Least Stable**
trans- disubstituted	*cis-* disubstituted	monosubstituted

Notes

Lecture Topic II.7: The E2 (Bimolecular Elimination) Mechanism

E2 is a Concerted Reaction to Produce an Alkene

We learned the general form of the E2 reaction in **Lecture Topic 1.8**. We will now study this mechanistic step is more detail.

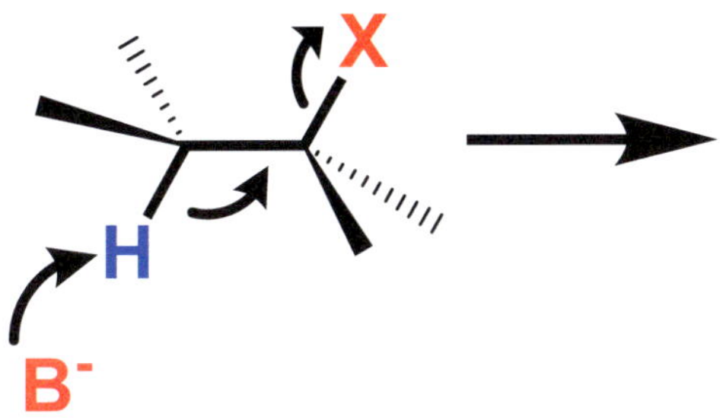

(A)

(B) Like the S_N2 reaction, the E2 reaction occurs in a single step. This type of reaction is called a:

(C) That there is only one step means that base, substrate and leaving group all influence rate. The rate law is:

A stronger base and better leaving group lead to faster rate.

Notes

Lecture Topic II.7: The E2 Mechanism

Solvent and Substrate Influences on E2 Rate

(A)

H-bonding solvents weaken Bases

A strong base is necessary for a rapid E2 reaction of an alkyl halide. H-bonding solvents (polar protic solvents) weaken base strength. Therefore:

(B)

It is also observed that more substituted alkyl halides react faster:

(C)

How can we use our knowledge of reactions to explain this observation?

Notes

Lecture Topic II.7: The E2 Mechanism
E2 Reactions Generally follow Zaitsev's Rule

More substituted alkyl halides yield more substituted alkenes

A

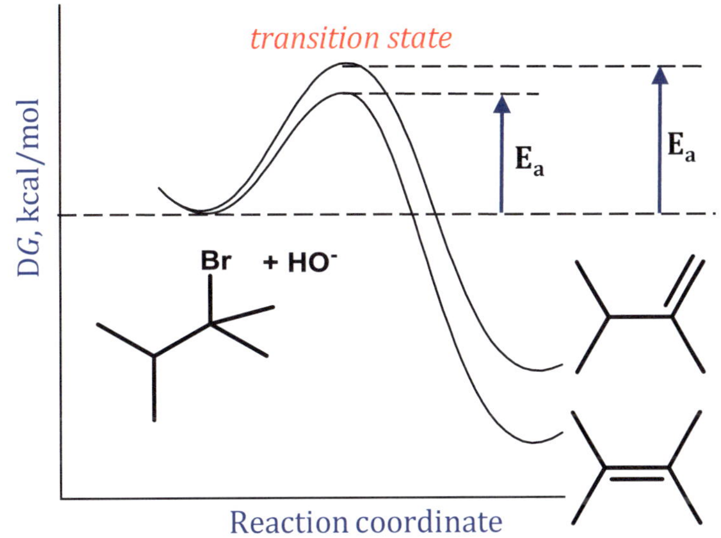

The E2 reactions generally follow "Zaitsev's Rule" like E1 reactions do.

Notes

Lecture Topic II.8: Stereoelectronic Effects in the E2 Reaction
Stereoelectronic Requirements

In order for the E2 reaction to work, the H and X to be eliminated must be **antiperiplanar**.

"Antiperiplanar" means the two groups are in the *anti-* conformation and in the same plane.

This is what is known as a **stereoelectronic** effect:

A

Notes

Lecture Topic II.8: Stereoelectronic Effects in the E2 Reaction

Influence on Mechanistically-Accessible Products

The stereoelectronic requirements of the E2 reaction must be considered when we attempt to predict products of a reaction.

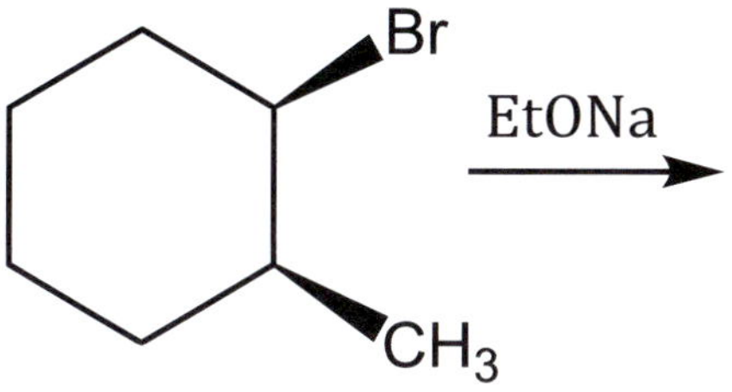

(A)

More stable alkene, major product

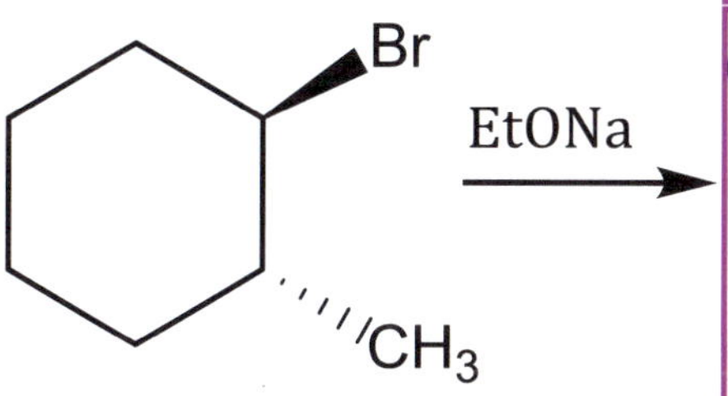

(B)

Less stable alkene, but no antiperiplanar H available for elimination to form more stable alkene. Note that the configuration about the chiral C is preserved because it was not involved in the reaction.

Notes

Lecture Topic II.8: Stereoelectronic Effects in the E2 Reaction
Required Conformation of Cyclic Substrate

For reaction on a cyclohexane derivative, eliminated groups must be axial to be antiperiplanar. If the eliminated H and leaving group are not axial in the most stable conformation, a ring flip may be necessary (may need to heat to accomplish this):

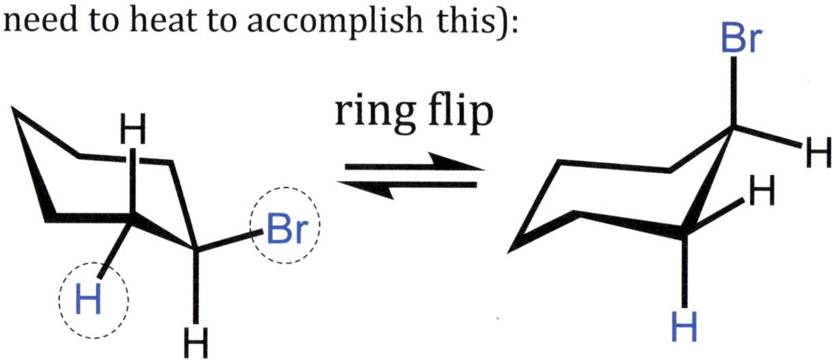

(A)

(B)

__Notes__

Lecture Topic II.8: Stereoelectronic Effects in the E2 Reaction
Conformational Analysis to Explain Reaction Rates

Example. The *cis-* isomer of 1-bromo-4-*t*-butylcyclohexane reacts by E2 reaction 500 times faster than does the *trans-* isomer. Why?

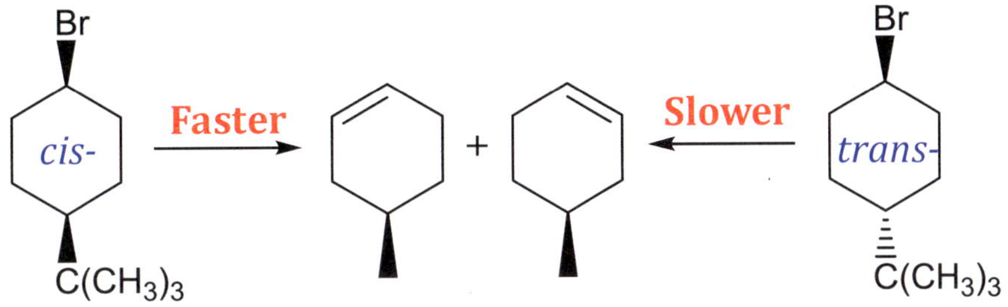

Start by drawing each reagent with Br (leaving group) in an axial position, where it must be for an E2 reaction to take place:

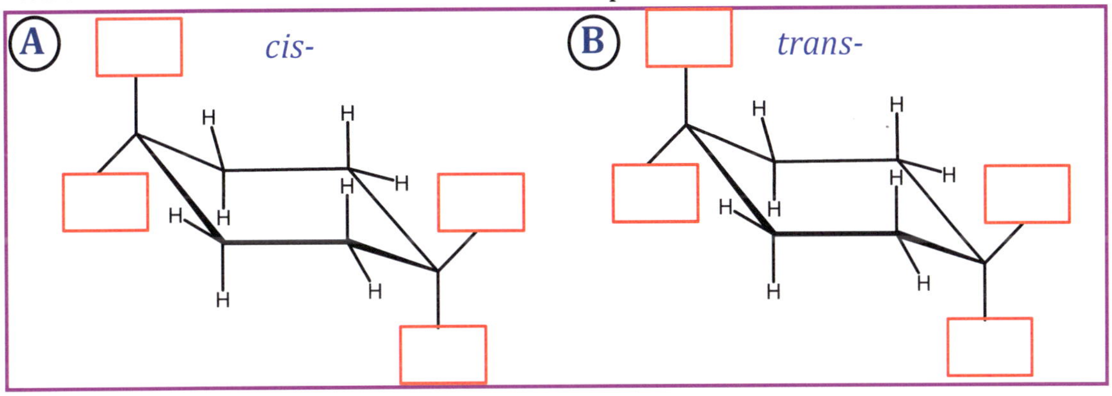

Notes

Lecture Topic II.8: Stereoelectronic Effects in the E2 Reaction
Conformational Analysis to Explain Reaction Rates

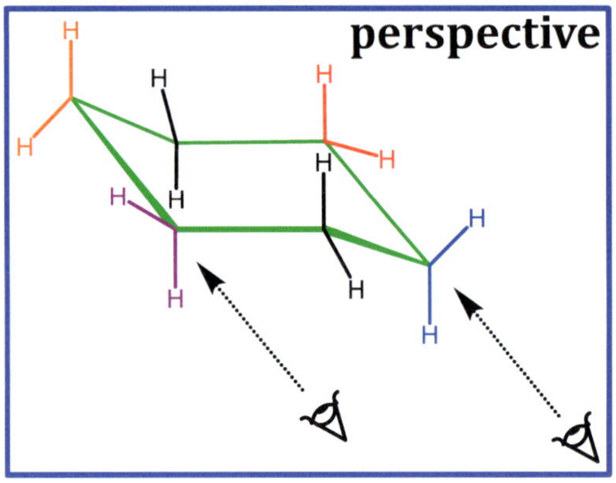

In the *cis*- conformation, when the leaving group is axial, the sterically encumbered *t*-butyl is equitorial. In the trans- conformation, both *t*-butyl and bromide can be equitorial. As a result, the *trans*- isomer reacts much more slowly (needs an energetically unfavorable ring flip to get Br axial to react!)

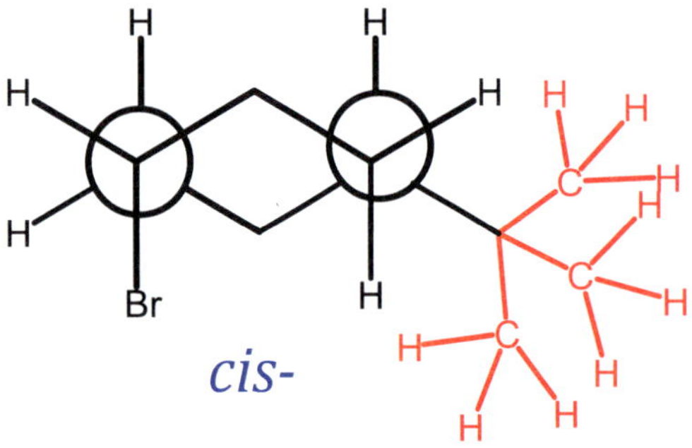

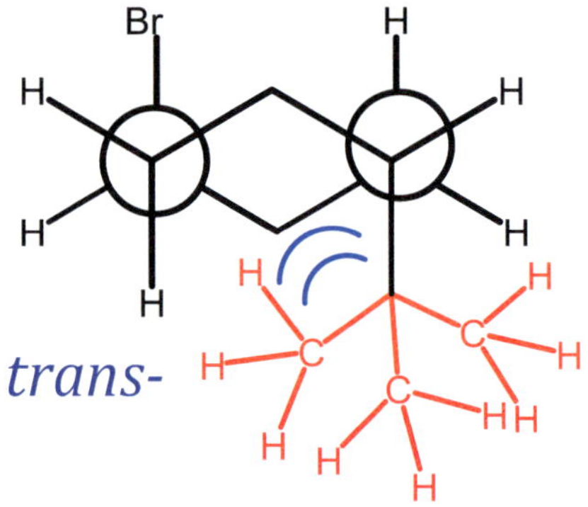

Notes

Lecture Topic II.8: Stereoelectronic Effects in the E2 Reaction

Substrate Configuration Influences Accessible Products

There are instances when only one H is present for elimination to form the most-substituted alkene. In these cases the antiperiplanar requirement may enforce formation of one alkene stereoisomer:

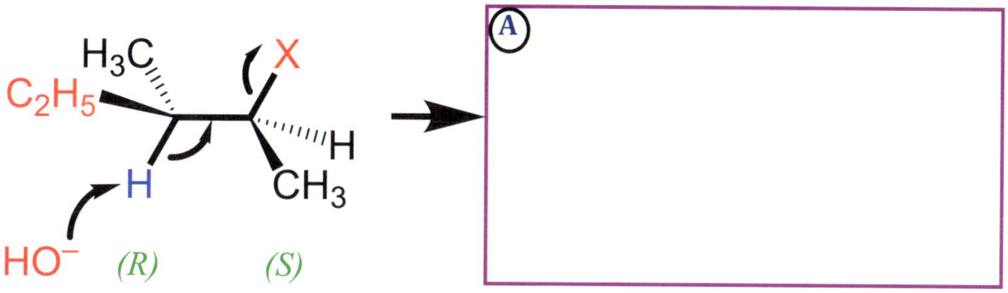

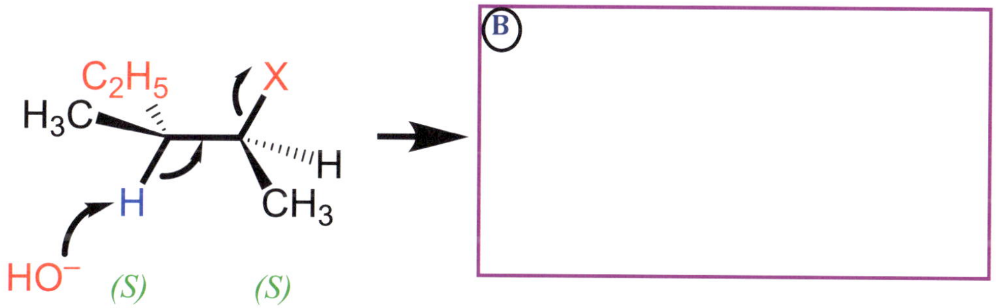

Notes

Lecture Topic II.9: Factors Leading to Non-Zaitsev Products in E2
Increasing the Amount of Non-Zaitsev Products

Steric hindrance to deprotonate a sterically-encumbered site increases as the base becomes bulkier:

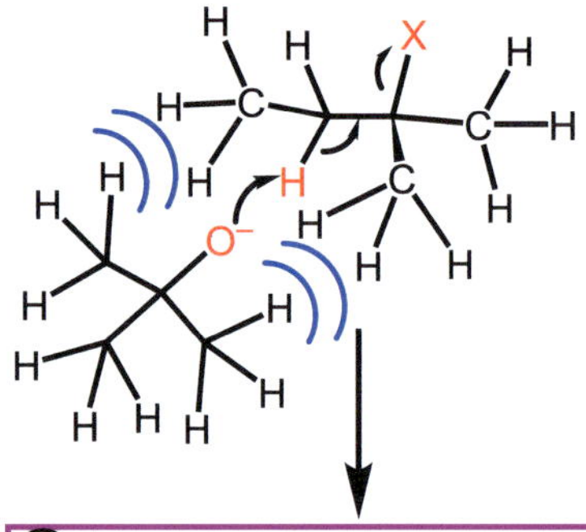

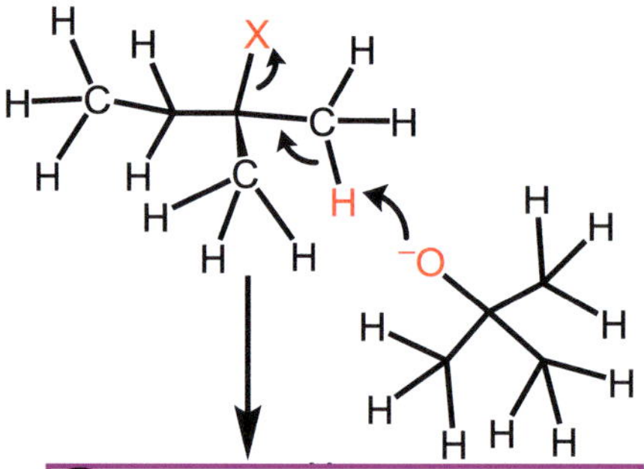

(A)

(B)

As a result:

(C)

Notes

Lecture Topic II.9: Factors Leading to Non-Zaitsev Products in E2

Hofmann Elimination: non-Zaitsev is the Major Product

The less-substituted product (non-Zaitsev) is sometimes called the

(A)

This terminology came from a clever version of the E2 reaction in which the leaving group is a bulky amine. This variation is called the:

(B)

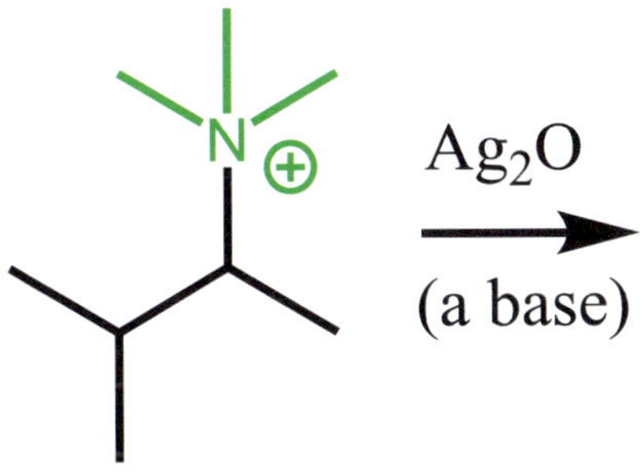

(C)

Conformational analysis may help us visualize and explain why the Hofmann product is favored...

Notes

Lecture Topic II.9: Factors Leading to Non-Zaitsev Products in E2
Hofmann Elimination: non-Zaitsev is the Major Product

On the left is the conformation necessary to produce the Zaitsev Product. On the right is the conformation necessary to produce the Hofmann Product.

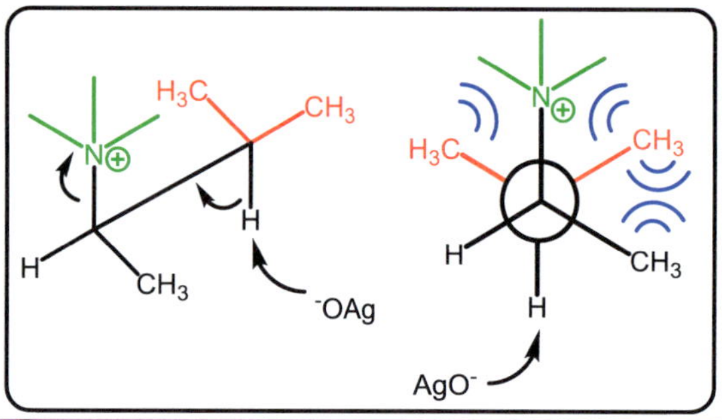

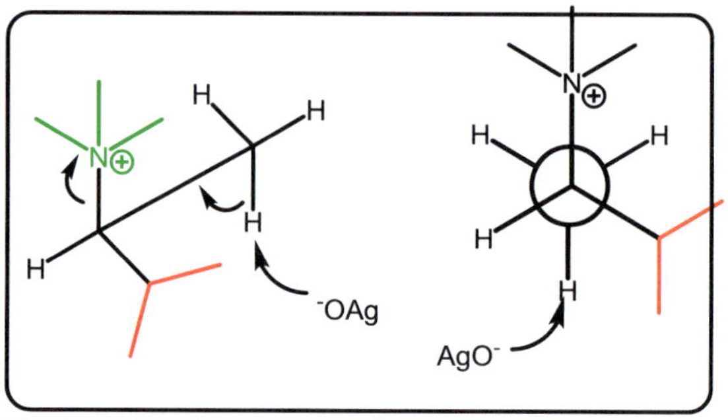

(A) Product:	(B) Product:
Disfavored because:	Favored because:

Notes

Lecture Topic II.10: Competition Among S_N1, S_N2, E1 and E2 Reactions

Predicting Major Products

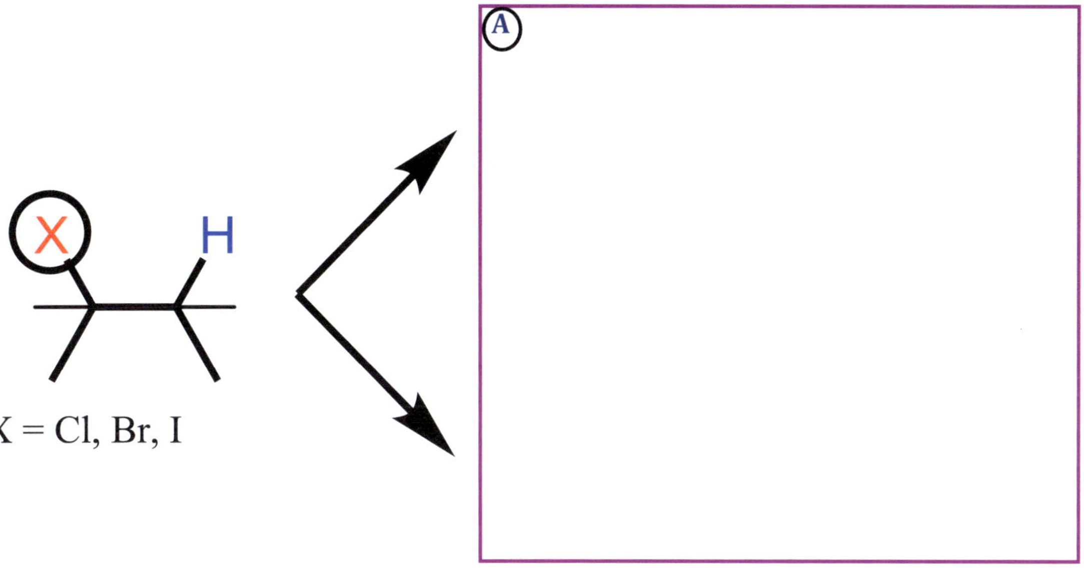

In a given context:

1. Elimination or Substitution?
2. E1, S_N1, E2, or S_N2?

Notes

Lecture Topic II.10: Competition Among S_N1, S_N2, E1 and E2 Reactions

Predicting Major Products

Bimolecular reaction (S_N2, E2) are favored by high concentration of good Nu or strong B

good Nu, weak B (i.e., I^-, Br^-, HS^-, NH_3): (A)

good Nu, strong B (i.e., HO^-, EtO^-, H_2N^-): (B)

poor Nu, strong B (i.e., $^tBuO^-$, bulky): (C)

Substrate can also help determine E2 versus S_N2

1° RX no bulky base: (D)

BULKY base (i.e., $^-O^tBu$): (E)

2° RX **S_N2 and E2**; the stronger/bulkier the base, the more **E2**

3° RX **E2 only**

Unimolecular reactions (S_N1, E1) are favored when neither a good nucleophile nor a strong base

Notes

Lecture Topic II.10: Competition Among S_N1, S_N2, E1 and E2 Reactions
Predicting Major Products

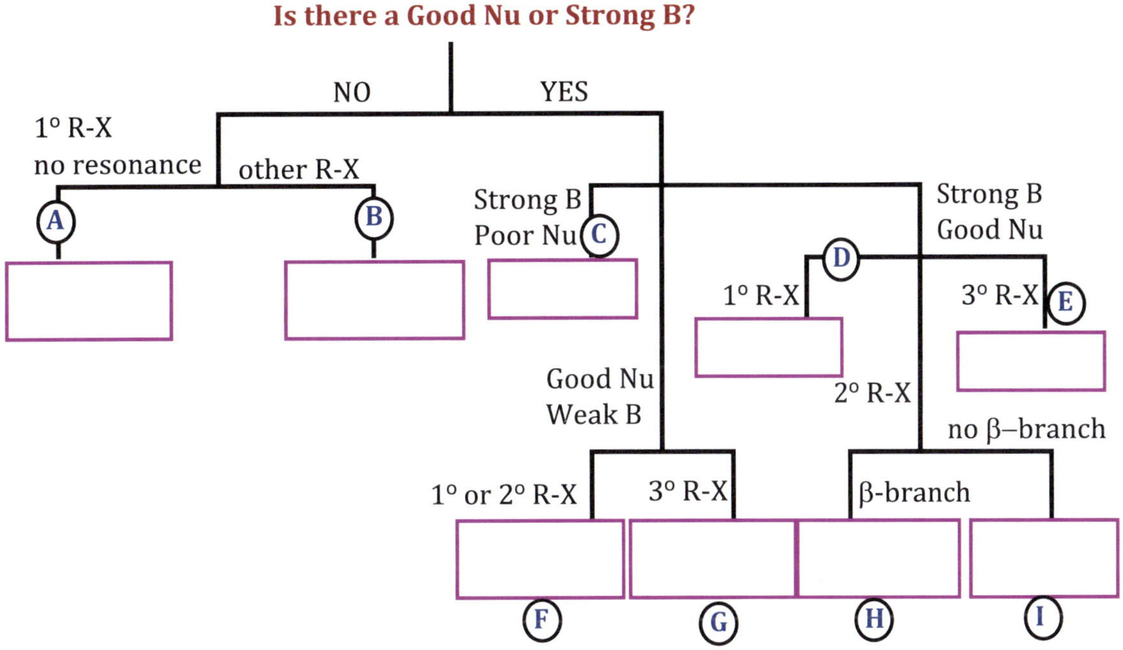

Notes

Lecture Topic II.10: Competition Among S_N1, S_N2, E1 and E2 Reactions

Predicting Major Products

Example. Determine whether each reaction will proceed predominantly via S_N1, S_N2, E1, E2, or some combination thereof, and show the product(s)

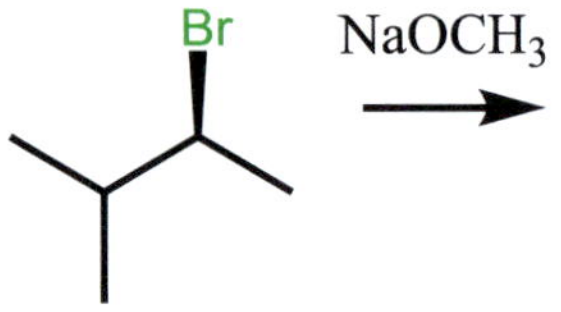

A

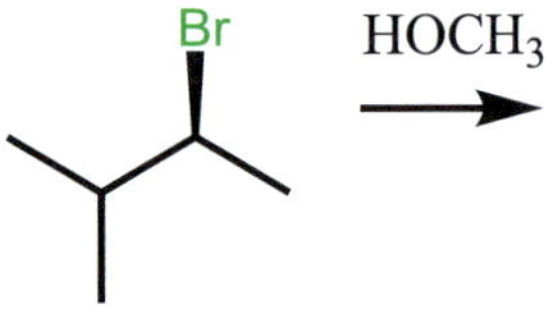

B

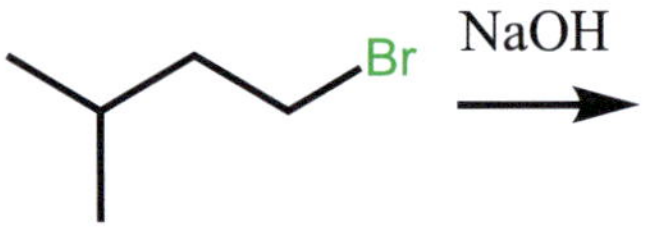

C

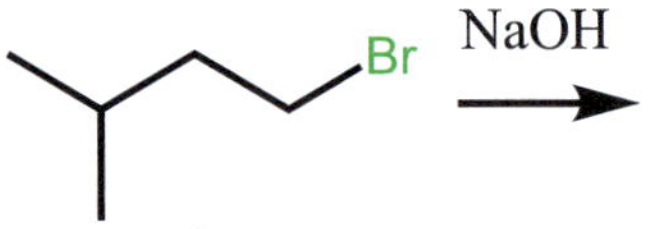

D

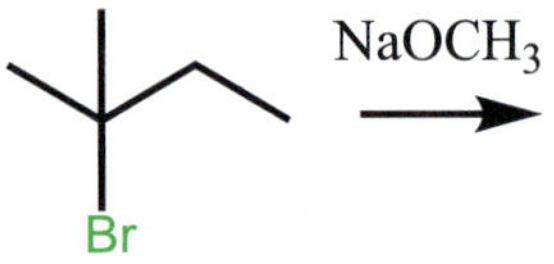

E

Notes

Lecture Topic II.10: Competition Among S_N1, S_N2, E1 and E2 Reactions
Predicting Major Products

Example. Predict whether each proceeds via S_N1, S_N2, E1, or E2, and draw the major product (show stereochemistry where applicable).

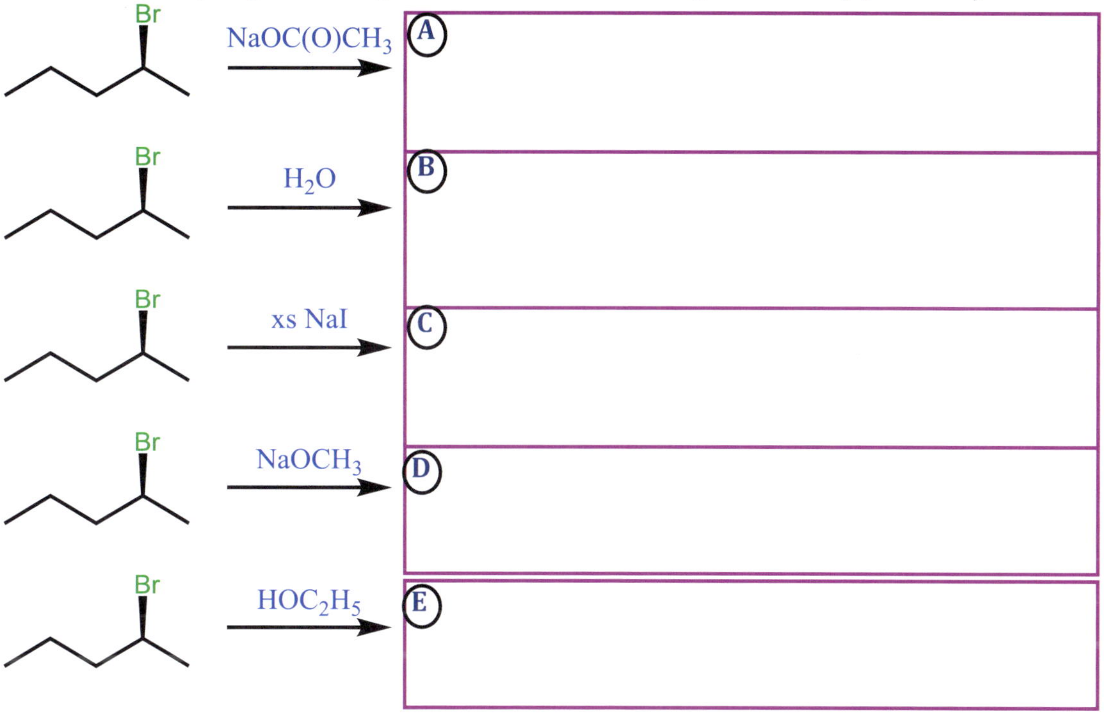

Notes

Lecture Topic II.11: Substitution and Elimination Reactions of Alcohols

Hydroxide is a Bad Leaving Group

So far we have only explored alkyl halides as substrates for S_N1, S_N2, E1 or E2 reactions because alkyl halides with Cl, Br or I substituents have the good leaving groups required. If we propose using an alcohol in these reactions:

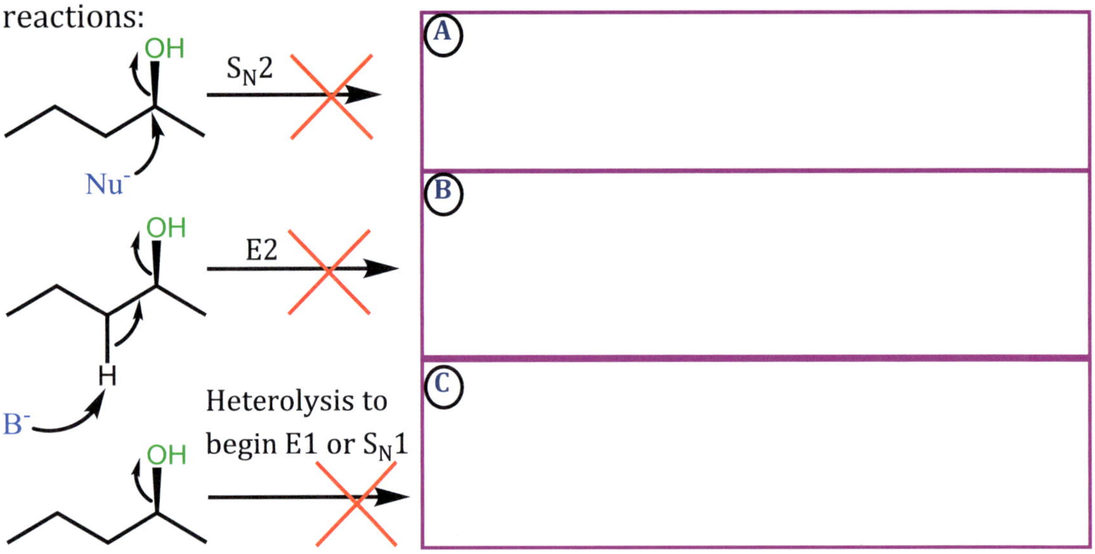

The hydroxide anion is a strong base (unstable anion) and is a bad leaving group. It must be **activated** prior to reaction by these routes.

Notes

Lecture Topic II.11: Substitution and Elimination Reactions of Alcohols

S_N1 Reaction of 3° ROH

One way to convert OH into a good leaving group is to protonate it with a strong acid. This leads to a water leaving group. This allows a tertiary alcohol to undergo an S_N1 reaction:

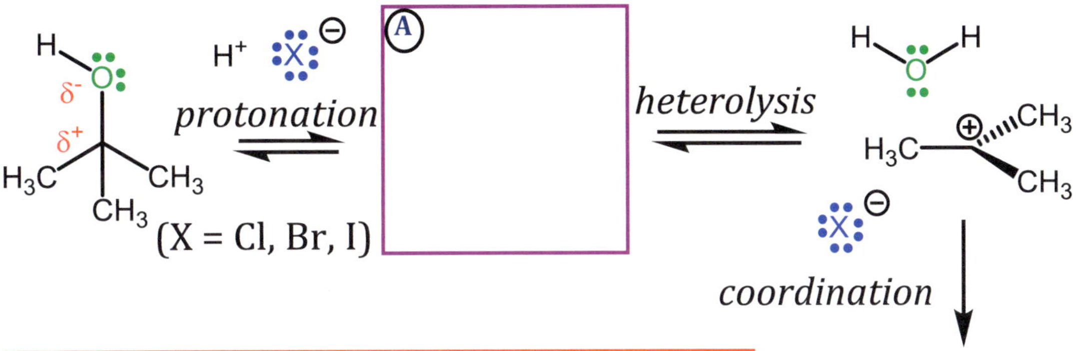

1. Activate OH to make it a good leaving group
2. S_N1 (heterolysis then coordination of X^-)

Everything we learned about S_N1 of alkyl halides applies here. Tertiary carbocation forms more rapidly, the carbocation can rearrange, etc.

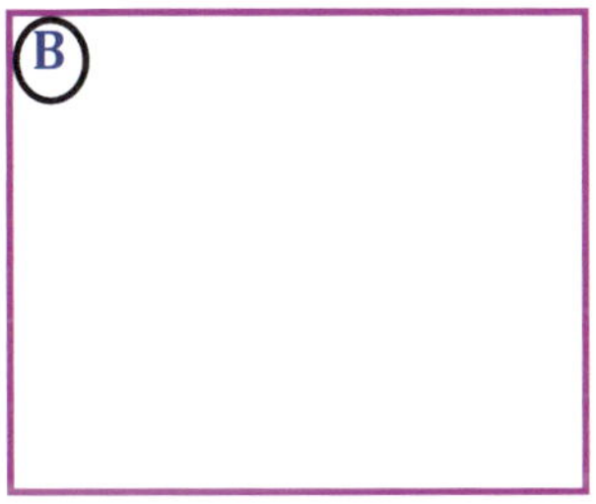

Notes

Lecture Topic II.11: Substitution and Elimination Reactions of Alcohols
S_N2 Reaction of Methyl, 1° and 2° ROH

An activated methyl, 1° or 2° alcohol will undergo S_N2 reaction with the halide, which is a good nucleophile:

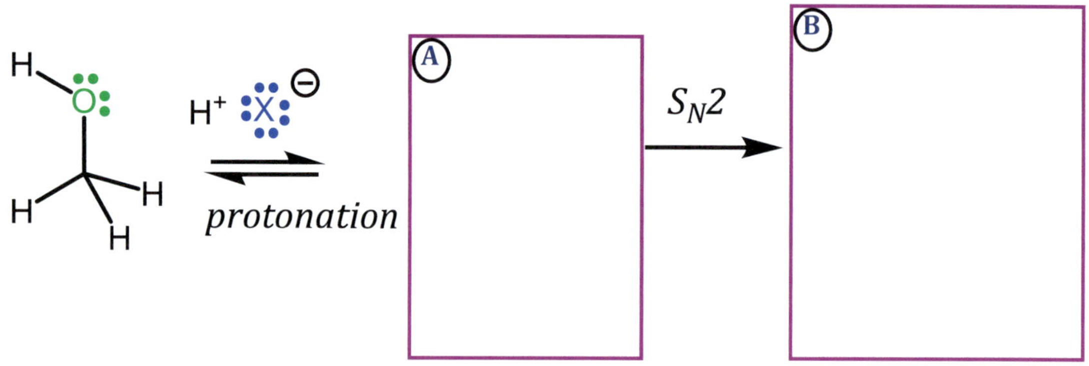

1. Activate OH to make it a good leaving group
2. S_N2 (concerted)

Everything we learned about S_N2 of alkyl halides applies here. Walden inversion occurs, less sterically-encumbered substrates react faster, etc.

Notes

Lecture Topic II.11: Substitution and Elimination Reactions of Alcohols
E1 of 2° and 3° ROH: Dehydration

If a 2° or 3° alcohol is activated with an acid having a non-nucleophilic counteranion (i.e., H_2SO_4 or H_3PO_4), an E1 Reaction will occur. Net reaction is loss of water from the alcohol, this particular type of E1 reaction is also called **dehydration**.

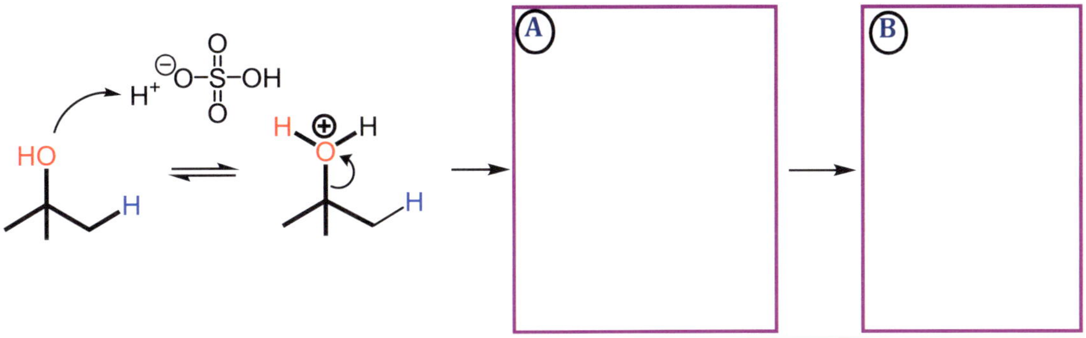

1. Activate OH to make it a good leaving group
2. E1 (heterolysis then electrophilic elimination)

Everything we learned about E1 of alkyl halides applies here. Tertiary carbocation forms more rapidly, the carbocation can rearrange, etc.
Note: cannot do E2 (needs strong **base**) of an alcohol with strong acid!

Notes

Lecture Topic II.11: Substitution and Elimination Reactions of Alcohols
Activation with Phosphorus Trihalide

Another way to activate an OH group is by reaction with PX_3 (X = Cl or Br):

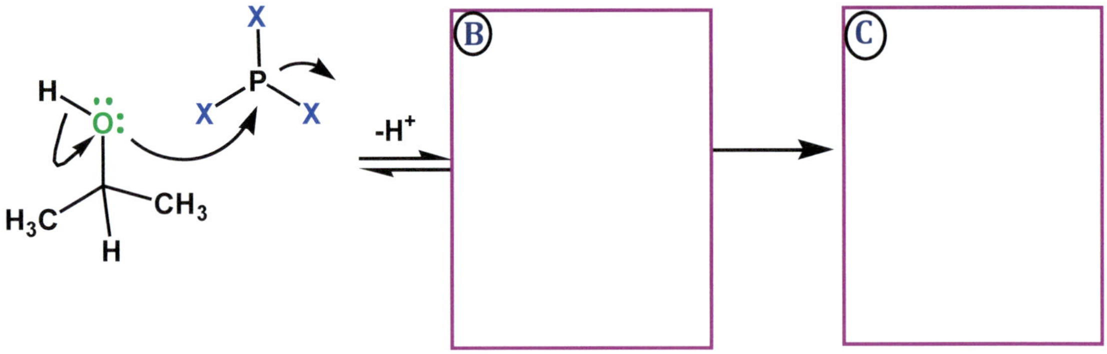

1. Activate leaving group
2. S_N2 reaction – all we have learned about S_N2 applies here; it will not work on a tertiary alcohol, for example.

Note: 1 mole PX_3 can produce 3 mol RX + H_3PO_3

Notes

Lecture Topic II.11: Substitution and Elimination Reactions of Alcohols
Activation with Thionyl Chloride

A third way to activate an OH group is by reaction with thionyl chloride:

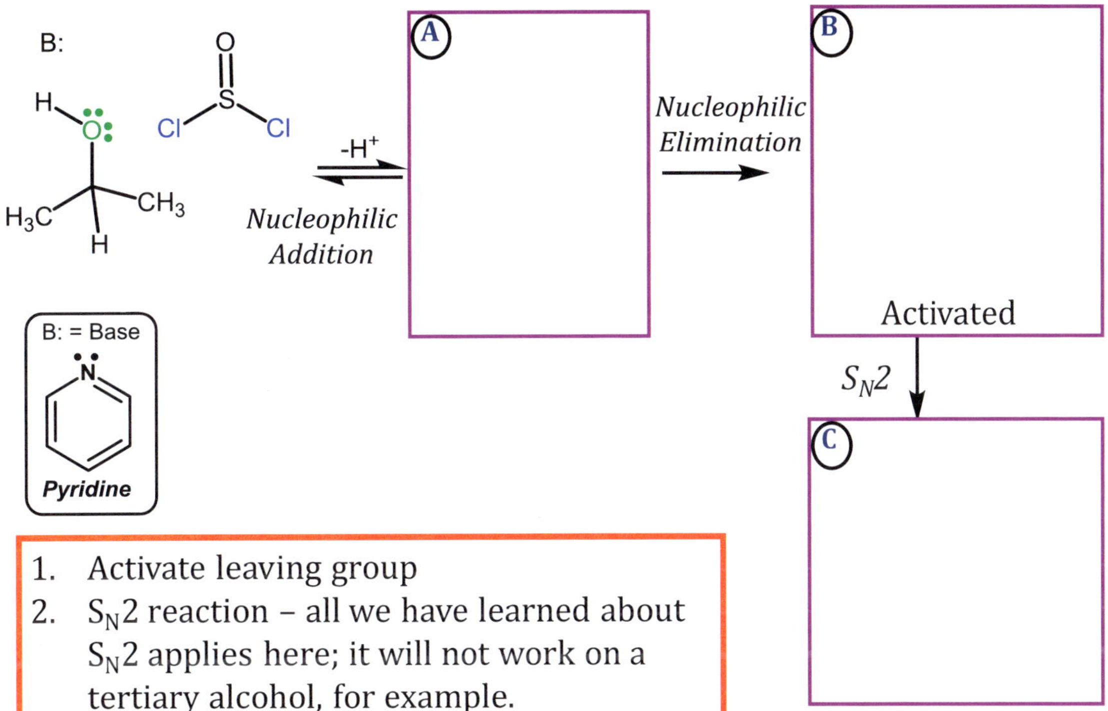

1. Activate leaving group
2. S_N2 reaction – all we have learned about S_N2 applies here; it will not work on a tertiary alcohol, for example.

Notes

Lecture Topic II.11: Substitution and Elimination Reactions of Alcohols

Conversion of ROH to Sulfonate Ester

Alcohols can also be converted into sulfonate esters. This is not a substitution or elimination reaction, but it is a good way to change the OH into a good leaving group:

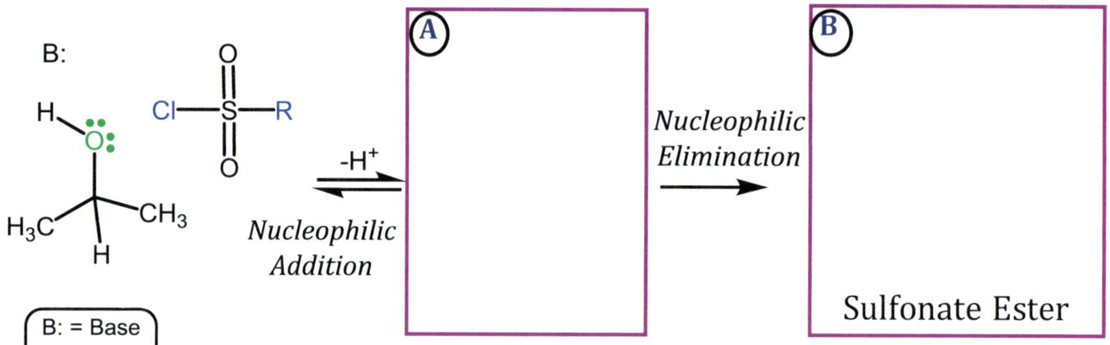

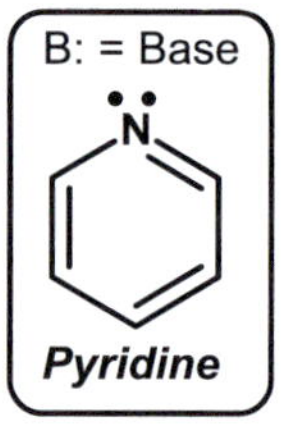

(C) Sulfonate Esters are useful because:

Notes

Lecture Topic II.11: Substitution and Elimination Reactions of Alcohols

Useful Sulfonate Esters

The sulfonate is a good leaving group because of resonance stabilization:

A sulfonate ester is ideal starting material for nucleophilic substitution:

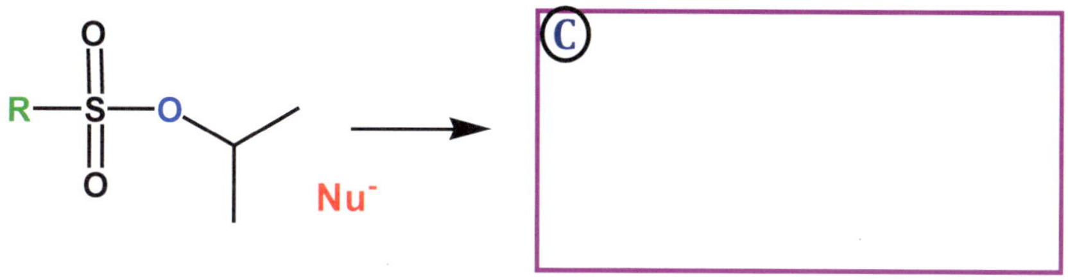

Three specific sulfonates are common:

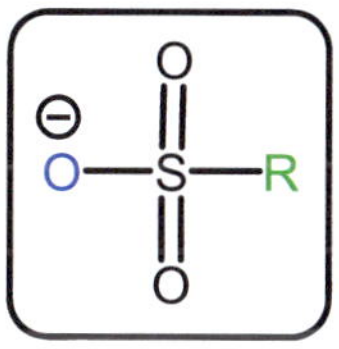

When R =	*Anion Name*
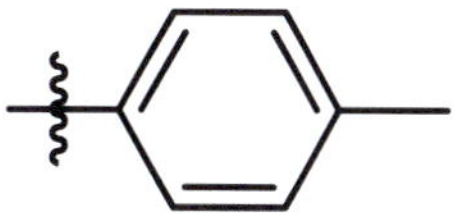	Tosylate
$-CH_3$	Mesylate
$-CF_3$	Triflate

Notes

Lecture Topic II.12: Oxidation and Reduction
Definitions for Organic Chemistry

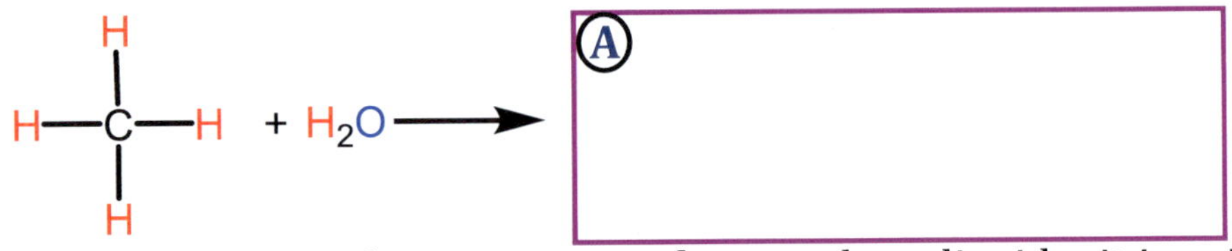

When methane burns in the presence to form carbon dioxide, it is said to undergo "oxidation". The opposite of oxidation is "reduction"

In general, the more C-H bonds that are replaced by C-O (or C to other more electronegative atom) bonds in an alkane, the more oxidized the carbon atom becomes:

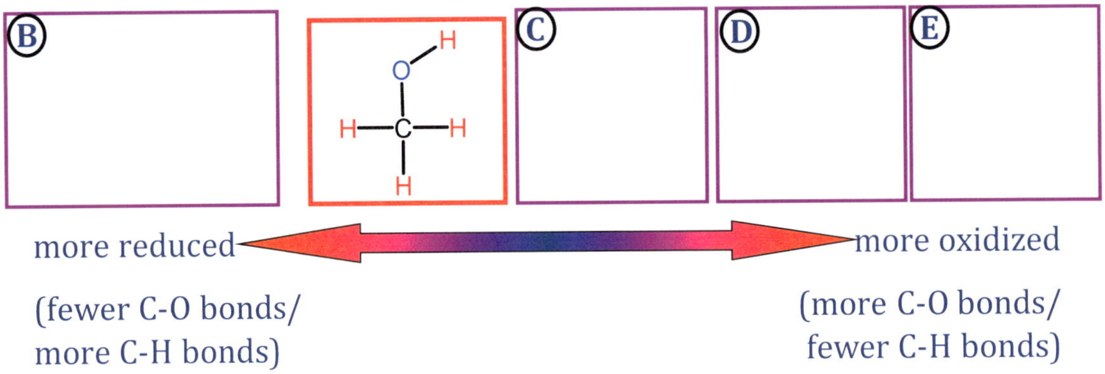

Notes

Lecture Topic II.13: Oxidation of Alcohols

Chromium Reagents are Commonly used to Oxidize Alcohols

Oxidation of alcohols is a useful way to make carboxylic acids, aldehydes, and ketones. Strong oxidizing agents like or H^+/CrO_4^{2-}, H^+/Cr_2O_7, CrO_3/H_2SO_4 (Jones Oxidation) replace all C-H bonds of an alcohol C with C–O bonds. PCC is weaker and can only replace one C–H with a C–O bond:

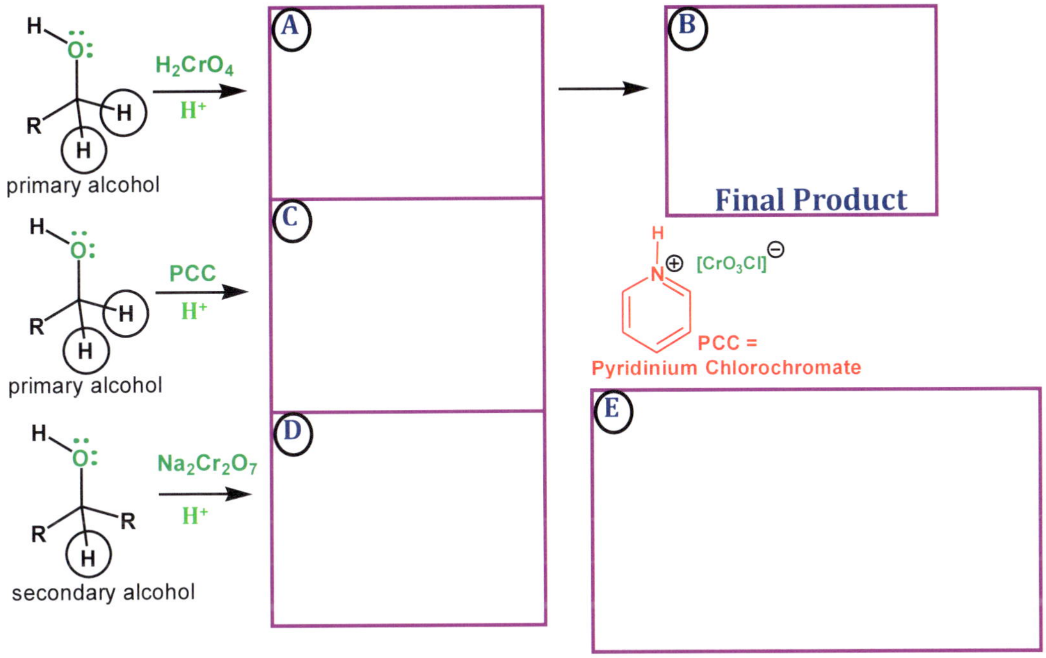

Notes

Lecture Topic II.14: Acid Cleavage of Ethers

S_N2 Reactions occur at Methyl and Primary Carbons

Like alcohols, ethers do not have a good leaving group for substitution reactions. However, the O atom of an ether can be protonated by HX (X = Cl, Br, I) to create a good leaving group. An S_N1 or S_N2 reaction follows:

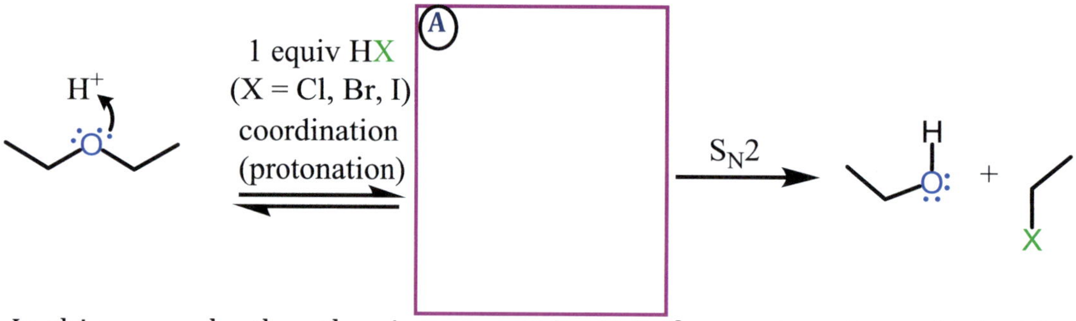

In this example, the ether is symmetric, so it does not matter which side we choose to attack with the nucleophile.

If the carbon to be attacked is methyl or primary:

(B)

Because:

Notes

Lecture Topic II.14: Acid Cleavage of Ethers

S_N1 Reactions occur at Secondary and Tertiary Carbons

If the carbon to be attacked by the nucleophile is 2° or 3°:

(A)

This is the case even though a good nucleophile (Cl-, Br- or I-) is present, so this **is a difference from alkyl halides**.

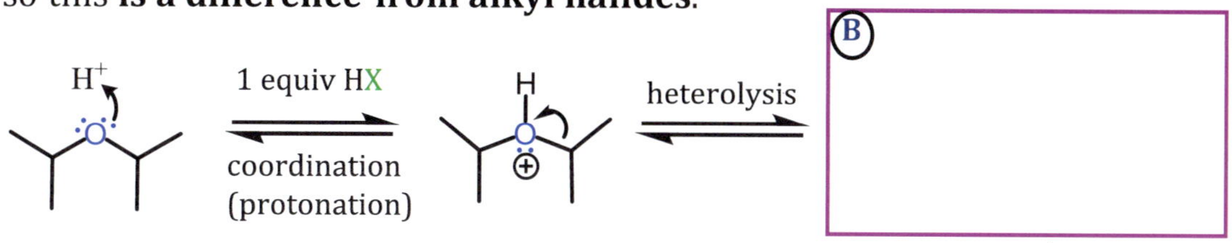

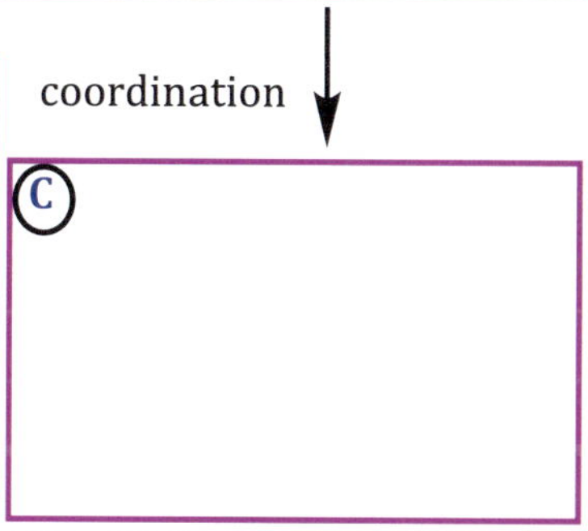

1. Activate OR to make it a good leaving group
2. S_N1 (heterolysis then coordination of X^-)

Everything we learned about S_N1 of alkyl halides applies here. Tertiary carbocation forms more rapidly, the carbocation can rearrange, etc.

Notes

Lecture Topic II.14: Acid Cleavage of Ethers

S_N1 is Preferred over S_N2 where Possible

The ether may not be symmetrical. In these cases:

(A)

If both sides of the ether are capable of S_N1, 3° reacts faster than 2°, as for any other S_N1 reaction. So:

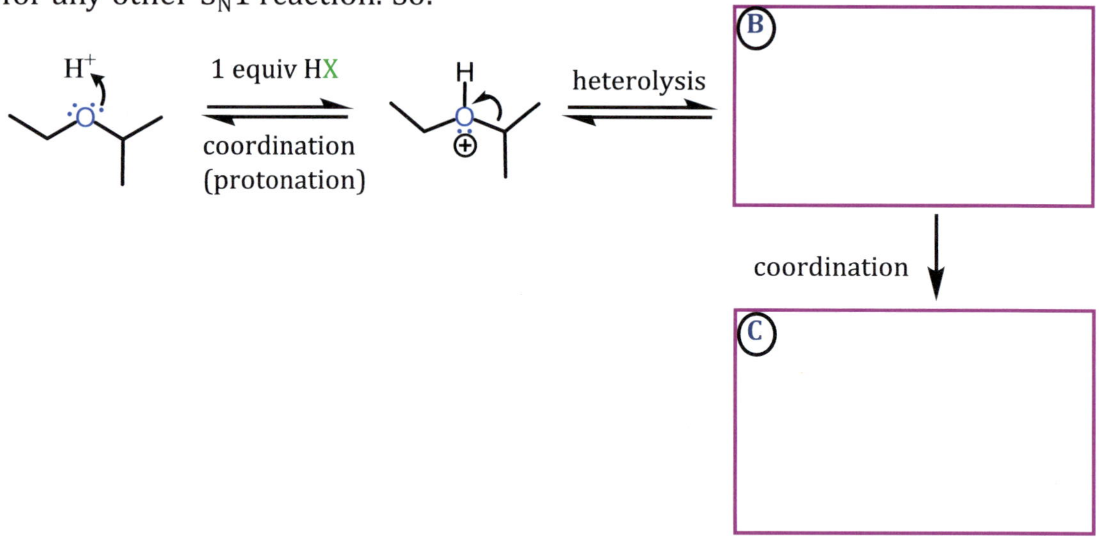

Notes

Lecture Topic II.15: Ring-Opening of Epoxides

Ring Strain Makes Epoxides More Reactive than other Ethers

An epoxides is a specific type of ether consisting of a 3-membered ring having an oxygen atom in the ring:

(A)

Epoxides are much more reactive than most other ethers because:

(B)

Nucleophiles can thus attack one of the electrophilic carbon atoms, alleviating the ring strain:

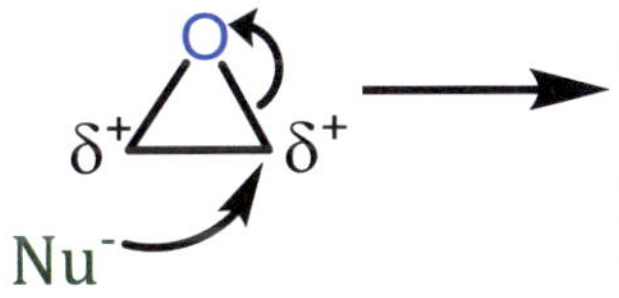

(C)

Notes

Lecture Topic II.15: Ring-Opening of Epoxides

Even a Poor Nucleophile will Attack a Cationic Species

Under acidic conditions, the epoxide oxygen is protonated just as in acid cleavage of other ethers:

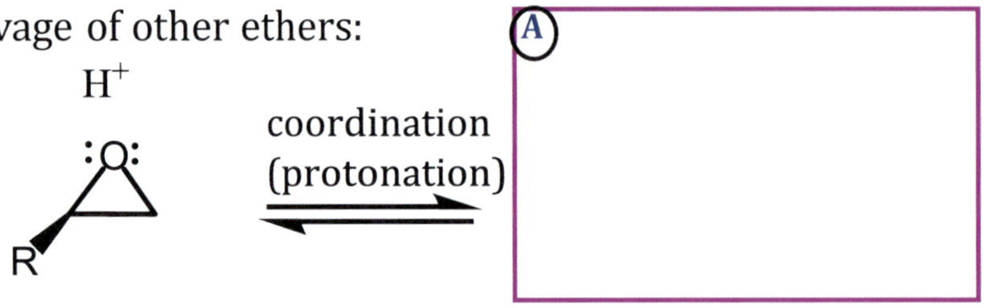

The more substituted C of the epoxide has more positive charge on it, so the nucleophile is selectively attracted to that site for an S_N2-like substitution:

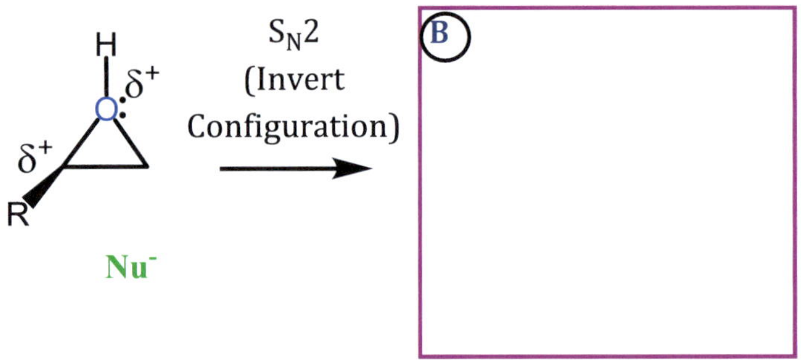

The cation has a strong pull for negative charge, so **even weak nucleophiles are reactive enough** to ring-open the epoxide under acidic conditions (H_2O, ROH, etc.)

Notes

Lecture Topic II.15: Ring-Opening of Epoxides

A Good Nucleophile is Needed to Attack a Neutral Species

Under basic (or simply non-acidic) conditions a typical S_N2 reaction occurs, which requires a good nucleophile:

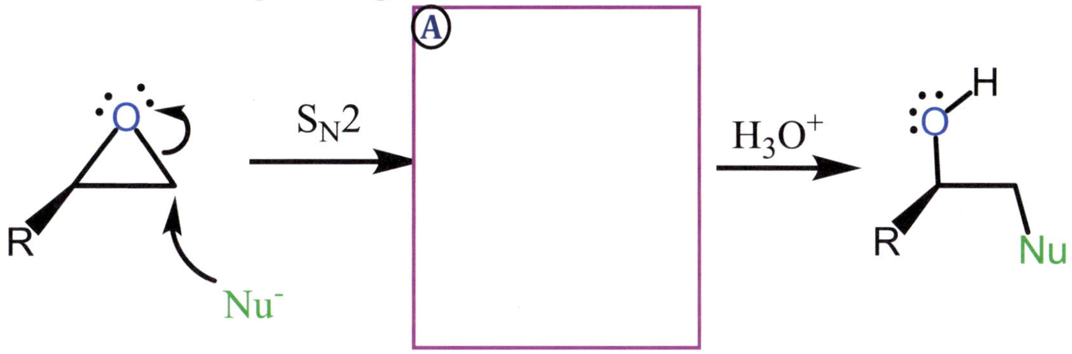

B Like other S_N2 reactions on neutral substrates, the nucleophile preferentially attacks:

Note that after ring-opening, the oxygen:

Notes

Example. Predict the major product of each reaction, showing stereochemistry where applicable.

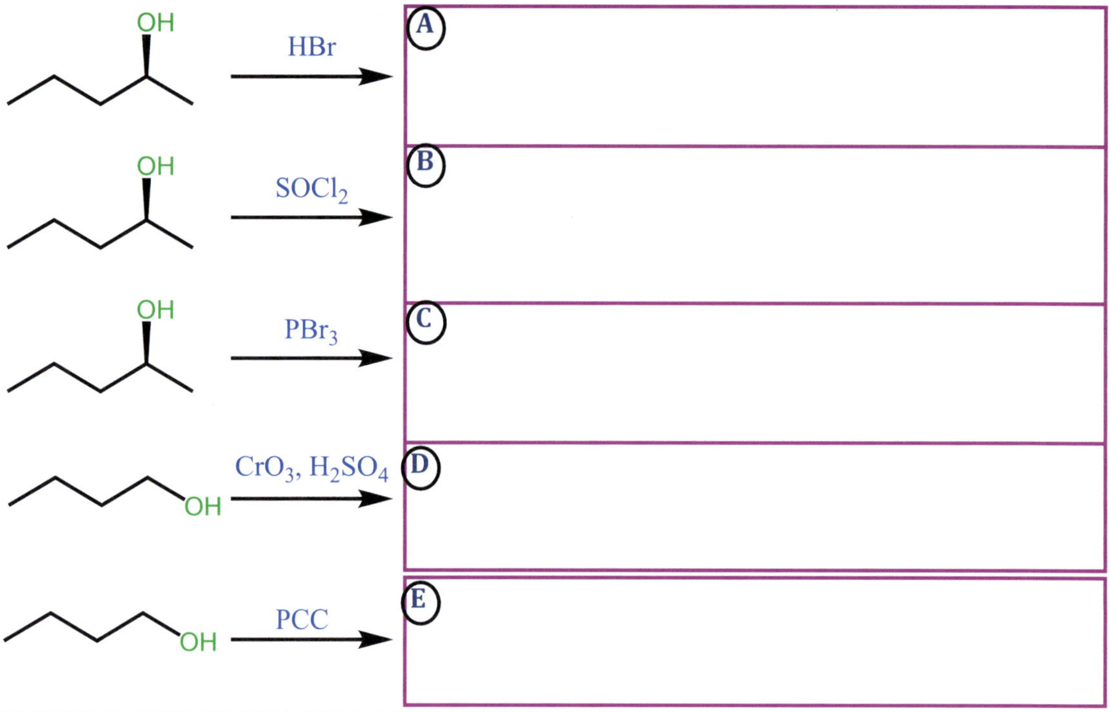

Notes

Lecture Topics II.11-15: Recap of Alcohol and Ether Reactivity

Substitution, Elimination and Oxidation

Example. Predict the major product of each reaction, showing stereochemistry where applicable.

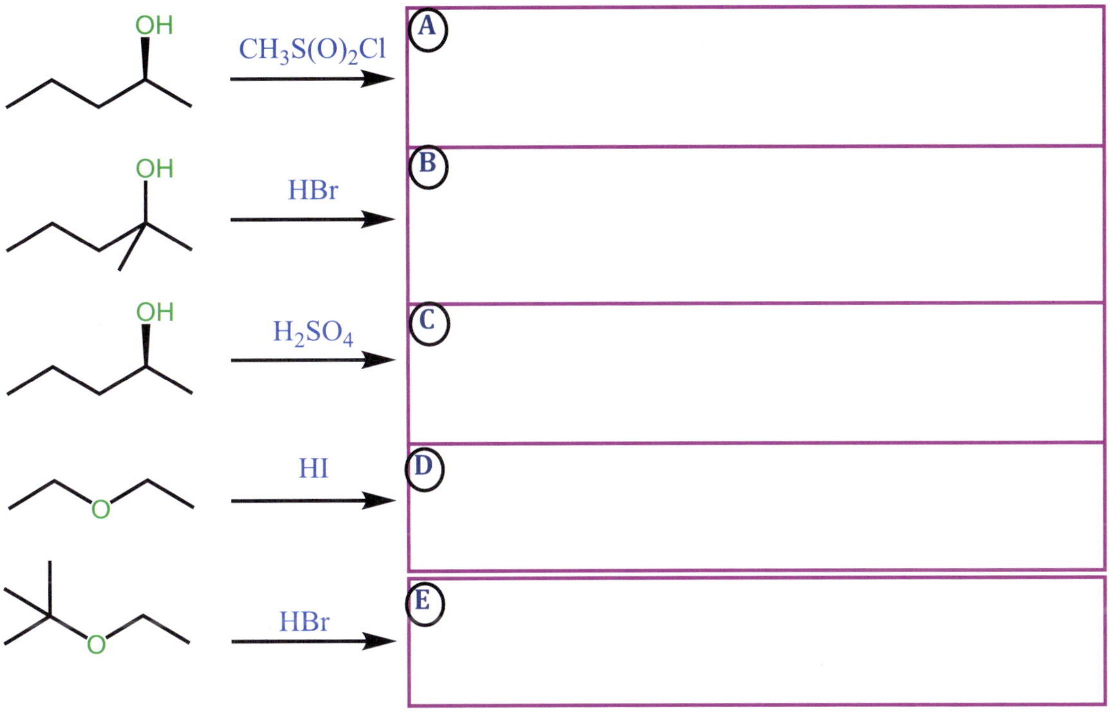

Notes

Example. Predict the major product of each reaction, showing stereochemistry where applicable.

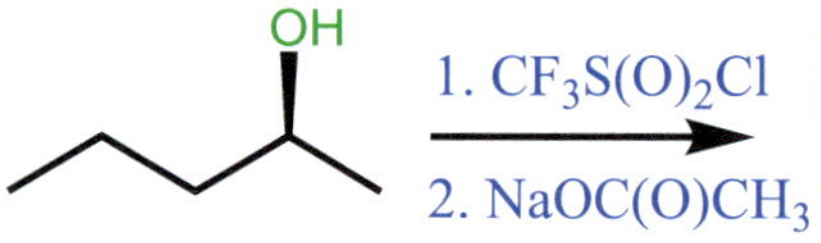

A

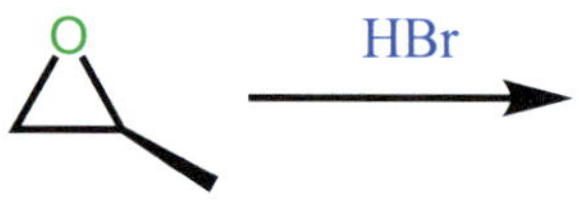

B

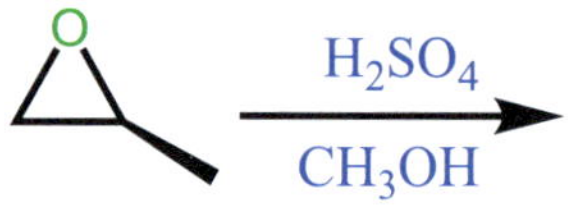

C

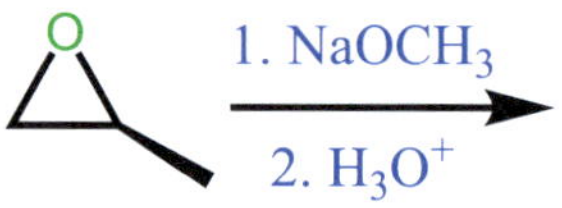

D

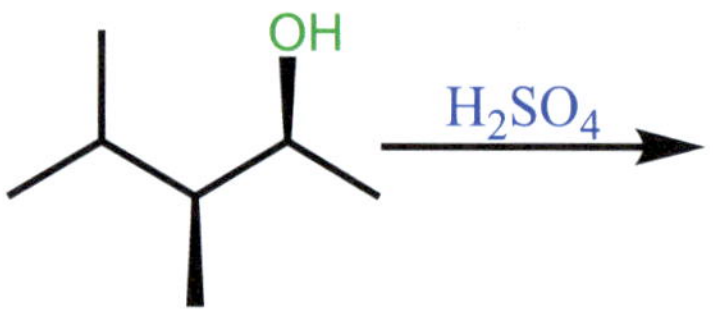

E

Notes

Part III: Reactions Involving Alkenes, Alkynes and Radicals

Lecture Topics in Part III:

1. Nomenclature II: Alkenes and the E-/Z- Convention
2. Properties of Alkenes and Hydrogenation of Alkenes
3. Electrophilic Addition to Alkenes and Carbocation Rearrangement Revisited
4. Markovnikov's Rule, Syn- and Anti- Addition
5. Reactions of Alkenes I: Hydrohalogenation and Hydration
6. Reactions of Alkenes II: Halogenation and Halohydrin Formation
7. Reactions of Alkenes III: Oxymercuration/Reduction
8. Reactions of Alkenes IV: Cyclopropanation and Epoxidation
9. Reactions of Alkenes V: Hydroboration/Oxidation
10. Reactions of Alkenes VI: Ozonolysis and Preparation of Vicinal Diols
11. Nomenclature III: Alkynes
12. Alkyne Reactions I: Hydrohalogenation and Halogenation
13. Alkyne Reactions II: Three Ways to Reduce Alkynes
14. Facile Constitutional Isomerism: Tautomerization
15. Alkyne Reactions III: Preparation of Carbonyls
16. Alkyne Reactions IV: Acetylide Anion Preparation and its use as a Nucleophile
17. Radicals I: Peroxide-Mediated Hydrobromination
18. Radicals II: Halogenation of Alkanes
19. Introduction to Polymers
20. Radicals III: Radical-Initiated Polymerization of Alkenes
21. Anionic and Cationic Polymerization of Alkenes

Lecture Topic III.1: Nomenclature of Alkenes

Naming Simple Alkenes

Nomenclature of alkenes is related to that for alkanes, with modifications:

1. always designate the longest chain having

(A) [ ] as the parent chain.

2. Number the parent chain such that the double bond

(B) [ ]

4. change the 'ane' ending of the name you would use if it were an alkane

(C) [ ]

5. Tell where the double bond starts:

(D) [ ]

Notes

Lecture Topic III.1: Nomenclature of Alkenes
Naming Simple Alkenes

 propene

 1-butene

 2-butene

 1-pentene

 2-pentene

 1-hexene

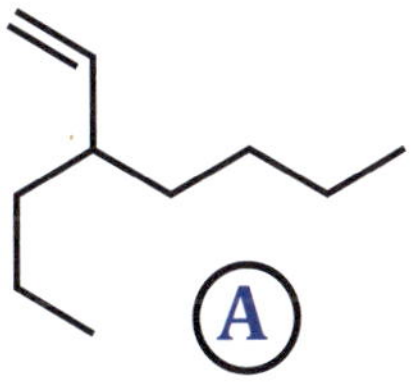

A

Notes

Lecture Topic III.1: Nomenclature of Alkenes
Alkenes with Multiple C=C or OH Units

6. If there is more than one double bond, use a “-diene”, “-triene”, etc. at the end of the name in place of the “ene”, with multiple numbers indicating where the double bonds are:

(A)

(B)

1,4,8-nonatriene

7. If an alcohol is present, it has higher priority than the, and the molecule has an “-ol” ending (with a number to tell you where the OH is!) after the “ene” suffix (change “ene” to “en”), and the **alcohol is given the lowest number**:

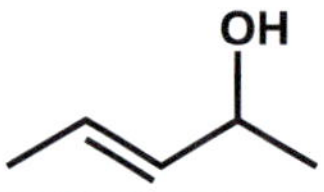

3-penten-2-ol

(C)

4-penten-1-ol

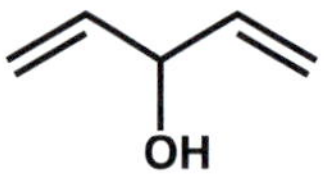

(D)

Notes

Lecture Topic III.1: Nomenclature of Alkenes
Configurational Isomerism in Disubstituted Alkenes

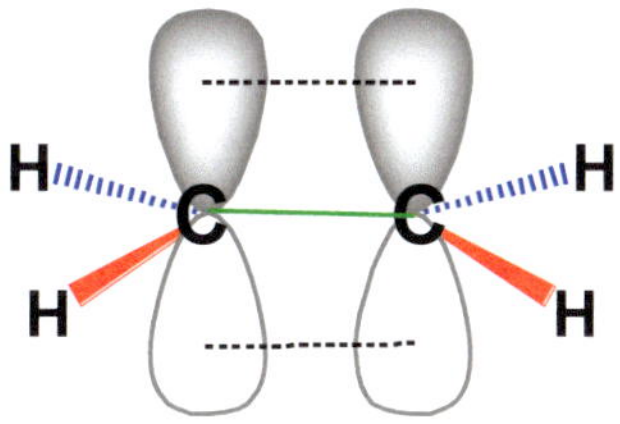

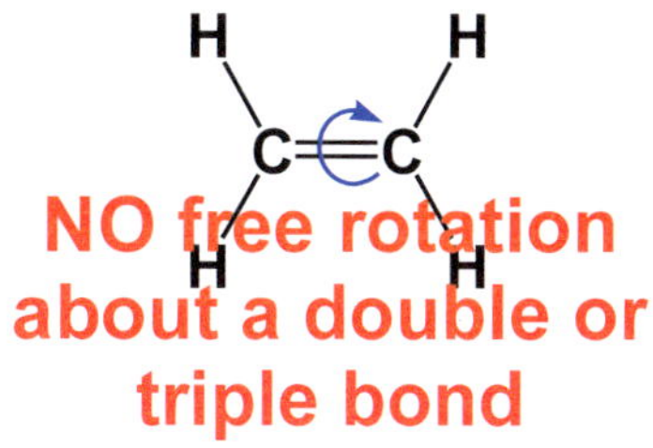

A

B

***cis*-alkene**

C

***trans*-alkene**

As we saw for cycloalkanes, when we have two configurational isomers, the one with two substituents on the same side is the *cis* isomer; the one with substituents on opposite sides is the *trans* isomer. The same holds for alkenes which have **TWO** non-hydrogen substituents.

Notes

Lecture Topic III.1: Nomenclature of Alkenes
Configurational Isomerism in Disubstituted Alkenes

Ⓐ 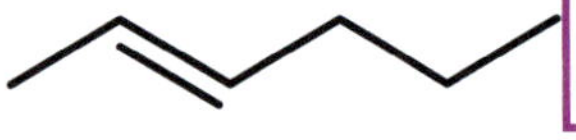 ☐-2-hexene

Ⓑ 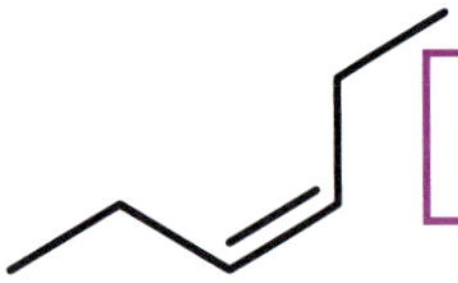 ☐-3-hexene

Ⓒ 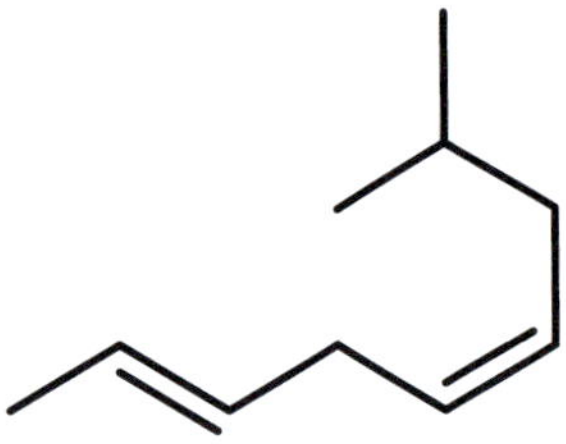 ☐-8-methyl-2,5-nonadiene

Ⓓ

Notes

Lecture Topic III.1: Nomenclature of Alkenes
What about Trisubstituted and Tetrasubstituted Alkenes?

Cis?

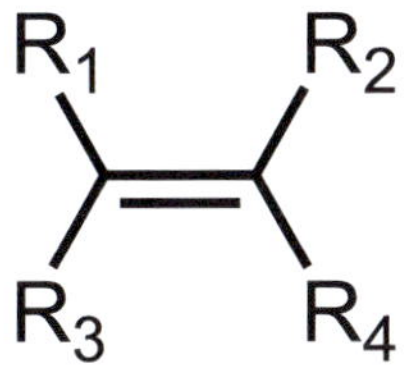

Trans?

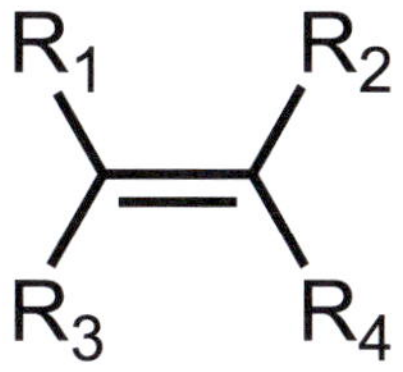

(A)

We must decide which is the "higher priority" substituent on each C of the C=C unit:

(B)

(C)

Notes

Lecture Topic III.1: Nomenclature of Alkenes
E- and Z- Labels of Configuration

For tri- or tetra-substituted alkenes we use *E*- (from German 'entgegen', opposite) or *Z*- (from German 'zusammen', together) notation. We prioritize units using CIP conventions (See **Topic I.20**)

If the two **highest priority groups** (are

on same side, then (A)

opposite sides, then (B)

(C) (D)

***Z*-isomer** ***E*-isomer**

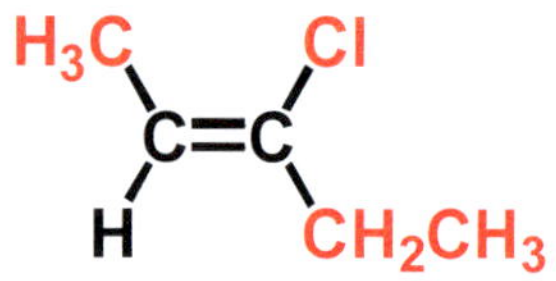

(E)

Notes

Lecture Topic III.1: Nomenclature of Alkenes
E- and Z- Labels of Configuration

Example. Designate each double bond as *E*- or *Z*-.

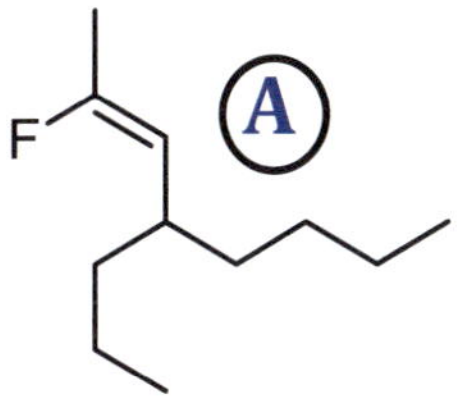

(A)

(B)

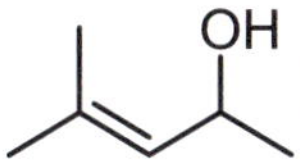

(C)

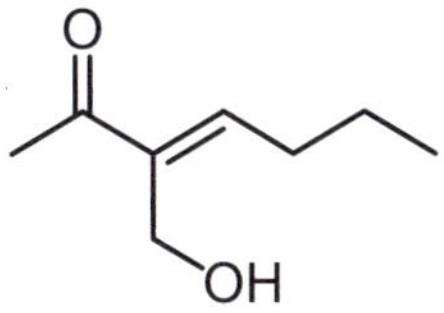

(D)

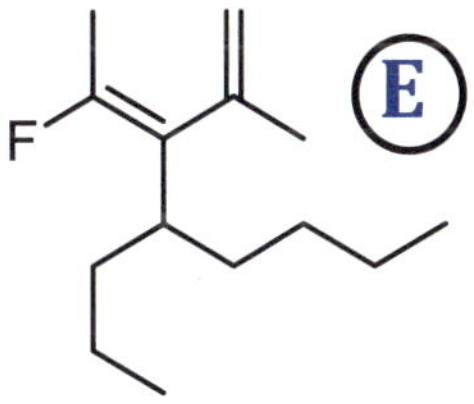

(E)

Notes

Lecture Topic III.2: Properties of Alkenes and Hydrogenation of Alkenes

Saturated versus Unsaturated Compounds

Alkanes: C_nH_{2n+2} **saturated** hydrocarbons
Alkenes: C_nH_{2n} **unsaturated** hydrocarbons (put a double bond in an alkane)
(olefins)

Each removal of two hydrogens from a saturated hydrocarbon (alkane) is one degree of unsaturation (can add one H_2 molecule to it)

A **degree of unsaturation** can be:

(A)

A general formula to determine the degree of unsaturation of a molecule from its formula ($C_cH_hN_nO_oX_x$, where X = F, Cl, Br or I) is:

(B)

*Note that oxygen's presence does not impact the degree of unsaturation

Notes

Lecture Topic III.2: Properties of Alkenes and Hydrogenation of Alkenes

Syn- and Anti- Addition to Pi Bonds

Alkenes can be prepared by elimination reactions. Alkenes can also undergo

(A)

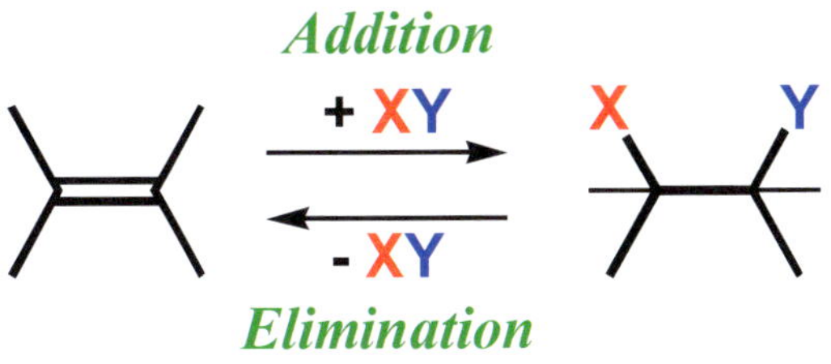

A stereochemical/mechanistic issue to be addressed is whether a given addition reaction occurs with the resultant substituents adding *syn-* or *anti-* to one another:

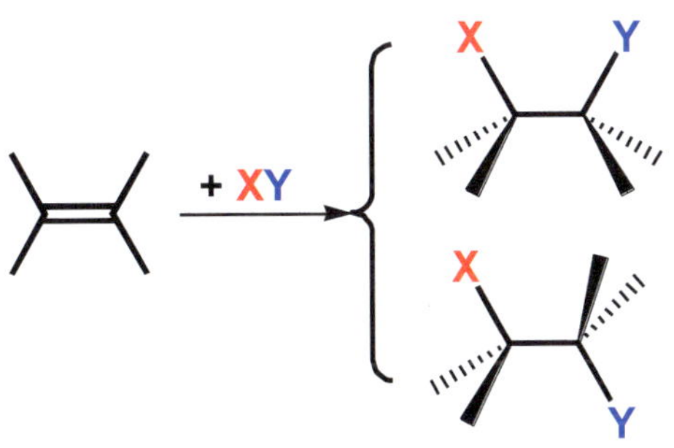

(B)

(C)

Notes

Lecture Topic III.2: Properties of Alkenes and Hydrogenation of Alkenes
Stability of Alkenes and the Hydrogenation Reaction

When we discussed Zaitsev's Rule (**Lecture Topic II.6**) we observed the following trend in alkene stability:

1. More substituted = More stable

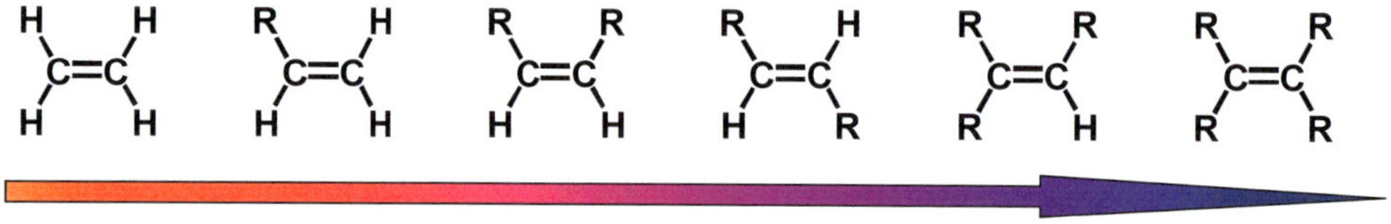

LESS STABLE MORE STABLE

These stabilities are determined by doing a **hydrogenation** of the alkene and measuring the heat given off by the process. Hydrogenation (***syn*- addition of two H atoms to the pi bond**) is accomplished by reacting with $H_2(g)$ and a Pd or Pt catalyst:

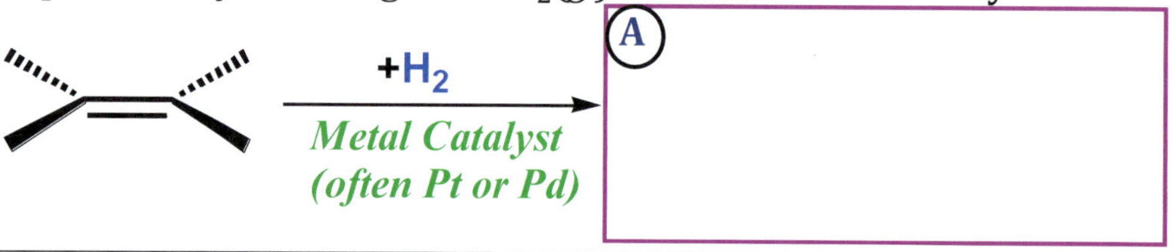

Notes

Lecture Topic III.2: Properties of Alkenes and Hydrogenation of Alkenes
Heats of Hydrogenation

Heat of Hydrogenation tells us about: (A) ______

A more stable alkene has a (B) ______

Higher energy!! (more heat given off)

Less Stable

Structure	heat of hydrog. (kcal/mol)
	30.1
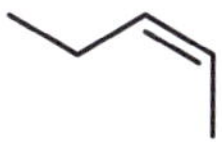	29.1
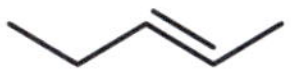	28.2
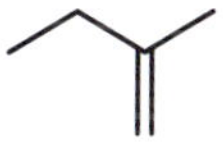	27.8
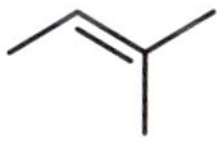	26.7

Lower energy!! (Less Heat Given Off)

More Stable

Notes

Lecture Topic III.3-5: Reactions Involving Electrophilic Addition of a Proton
Hydrohalogenation

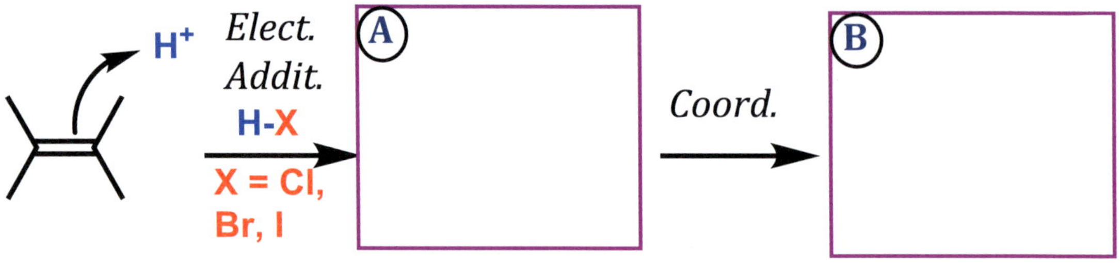

The mechanism is the reverse of (C)

The rate limiting step is

(D)

The carbocation can: (E)

Addition is:

(F)

Notes

Lecture Topic III.3-5: Reactions Involving Electrophilic Addition of a Proton
The More Stable Intermediate forms More Rapidly

If the two pi-bonded carbon atoms are differently-substituted, there will be two choices for carbocation formation upon electrophilic addition:

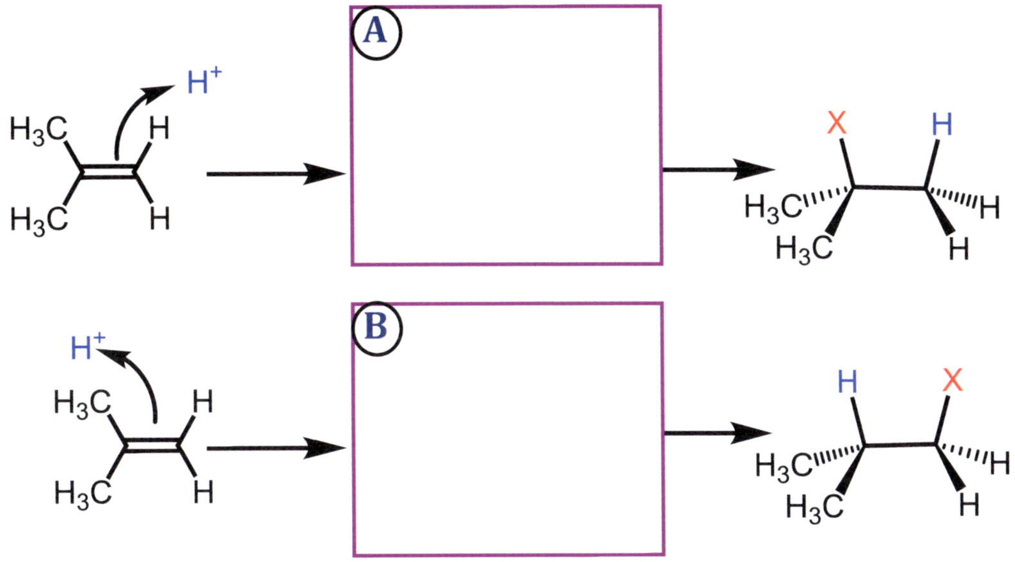

The **major product** is that derived from the more stable carbocation, in this case the tertiary cation.

Notes

Lecture Topic III.3-5: Reactions Involving Electrophilic Addition of a Proton
"Markovnikov's Rule"

OBSERVATION= Markovnikov's rule:

(A)

EXPLANATION=

(B)

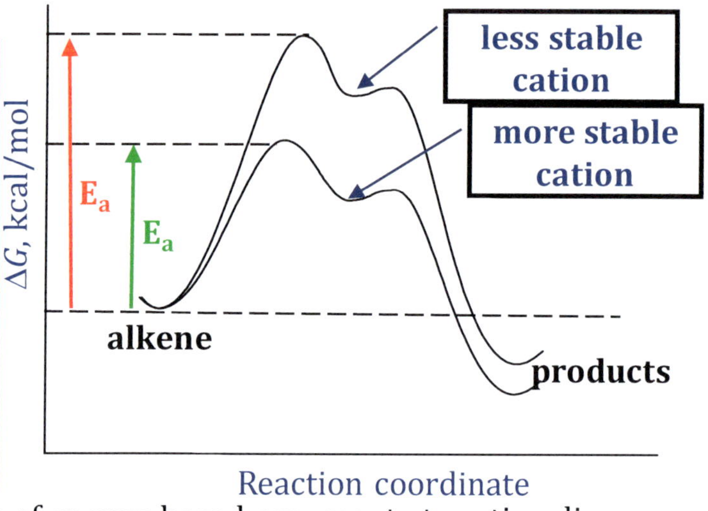

We can use the same type of energy-based arguments to rationalize Markovnikov's rule as we did to rationalize Zaitsev's rule.

Notes

Lecture Topic III.3-5: Reactions Involving Electrophilic Addition of a Proton
Carbocations can Rearrange to More Stable Species

Remember that any time a carbocation is formed in the course of a reaction, you will need to assess whether it rearranges.

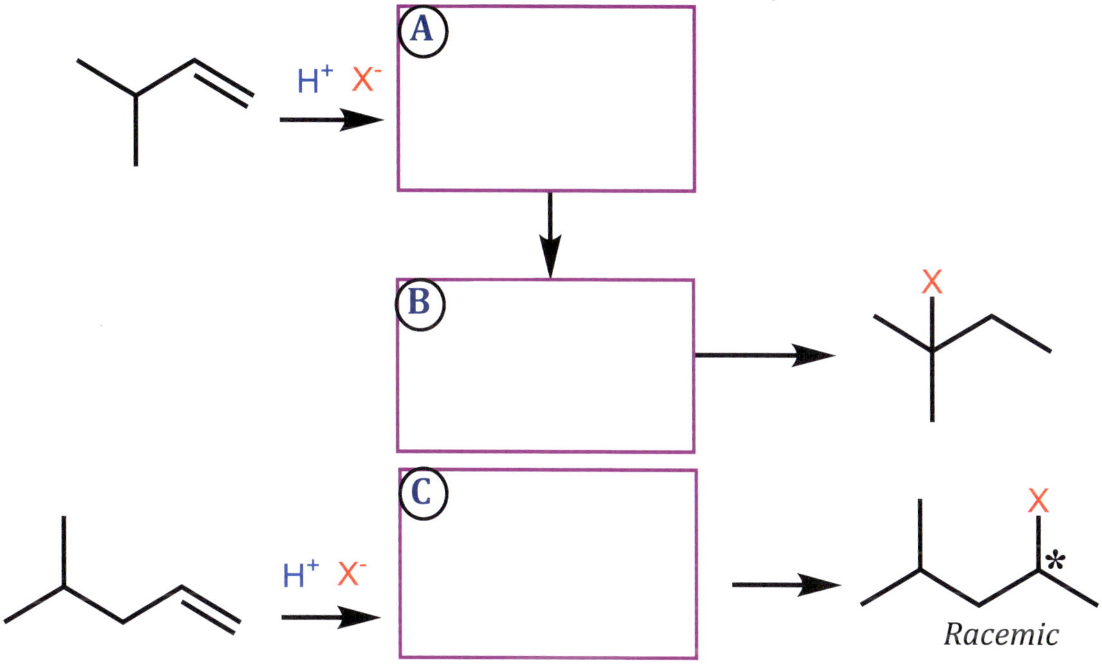

Notes

Lecture Topic III.3-5: Reactions Involving Electrophilic Addition of a Proton
Hydration

Addition of HX to an alkene is the reverse of E1 on an alkyl halide. We can also do a **hydration reaction** (addition of water to an alkene), which is the reverse of the dehydration (E1) of alcohols:

NET REACTION:

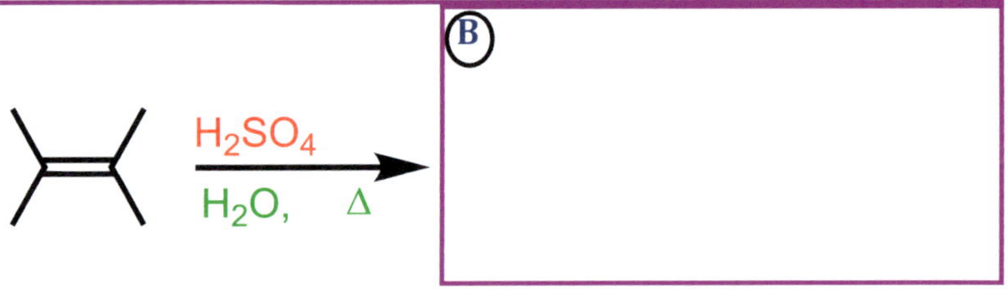

Individual Steps:

This reaction gives the Markovnikov product and has 50:50 syn-anti addition. The carbocation can rearrange just as for hydrohalogenation.

Notes

Lecture Topic III.6: Halogenation and Halohydrin Formation

Formation of a Halonium Intermediate

When a halogen (Cl_2 or Br_2) comes in contact with an alkene, it sets off a series of events:

1. C=C pi electrons repel 'soft' electron cloud of halogen

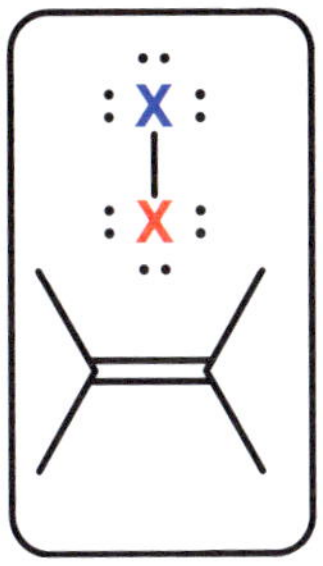

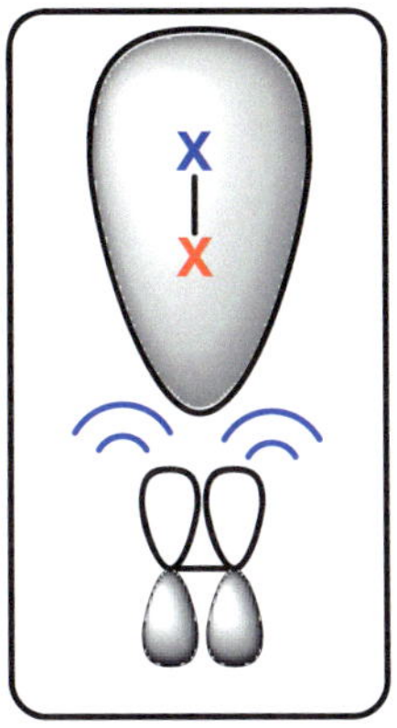

2. This causes an induced dipole in the halogen:

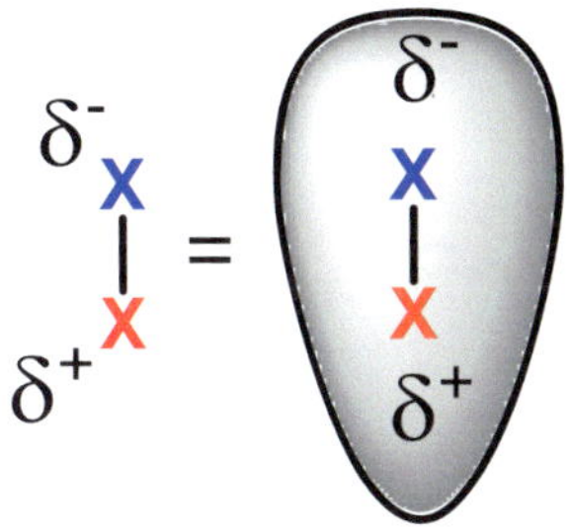

The partially positive end of the induced dipole now has an attraction for the nucleophilic alkene pi bond electrons ...

Notes

Lecture Topic III.6: Halogenation and Halohydrin Formation

Formation of a Halonium Intermediate

3. The polarized halogen begins an electrophilic addition to the alkene, losing X^{-}:

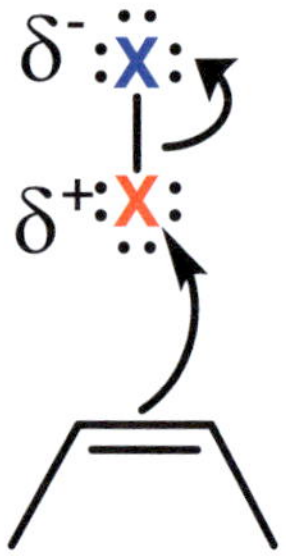

4. As the positive charge begins to form on the carbon, a halogen lone pair is naturally attracted to it:

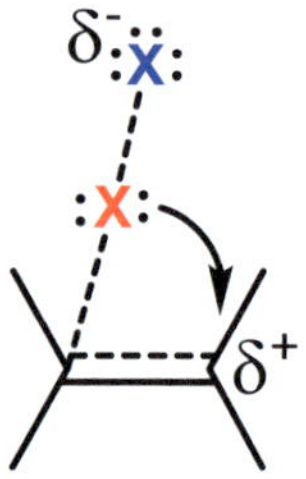

So that the final arrow pushing to give the resultant **halonium intermediate** is:

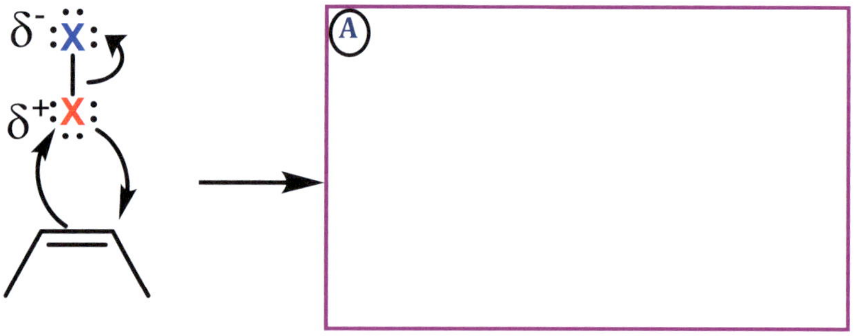

Notes

Lecture Topic III.6: Halogenation and Halohydrin Formation

Formation of a Halonium Intermediate

The formation of the halonium intermediate (chloronium if X = Cl, bromonium if X = Br) is concerted, so the relative distribution of substituents in the alkene is maintained. Keep in mind the stereochemistry!

achiral

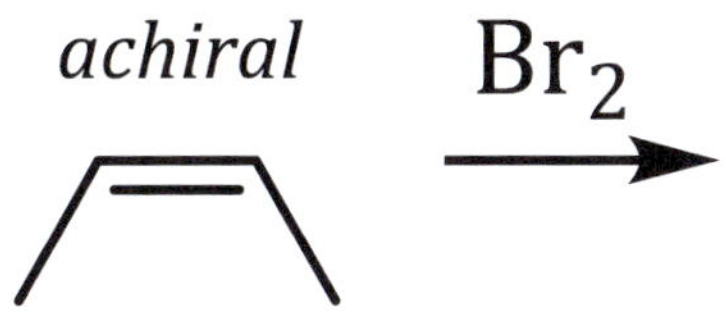

Meso Compound

achiral

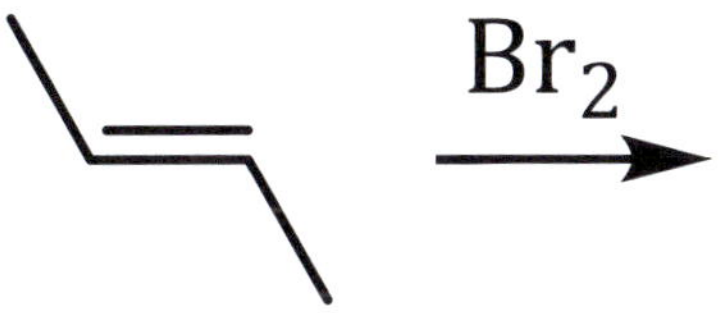

Racemic

Notes

Lecture Topic III.6: Halogenation and Halohydrin Formation

Reactivity of a Halonium Intermediate

Once formed, the halonium intermediate reacts very much like the oxonium intermediate formed by protonating an epoxide. This is not surprising since they are both three-membered ring cations.

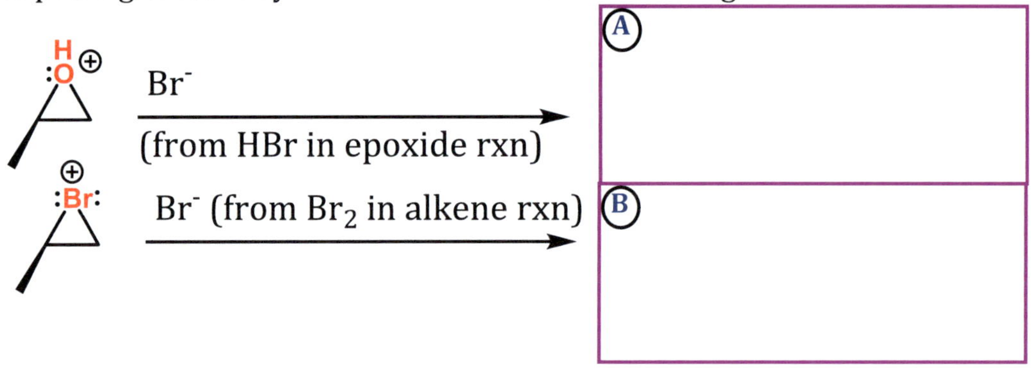

The net product of the halogenation of an alkene with Br_2, then, is:

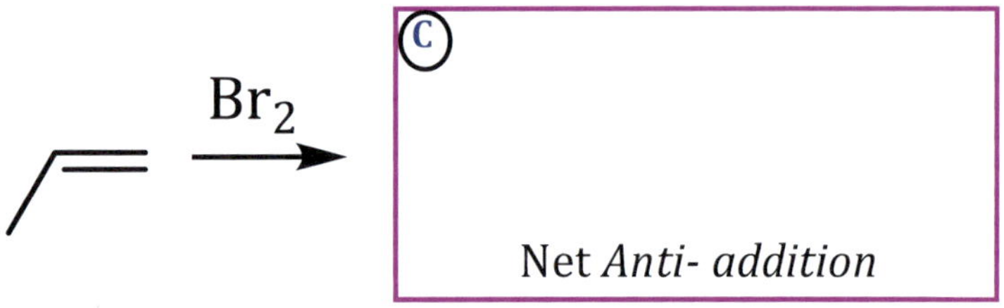

Notes

Lecture Topic III.6: Halogenation and Halohydrin Formation

Reactivity of a Halonium Intermediate

If an excess of water is present when the halonium intermediate forms, then the H_2O will serve as a nucleophile. A species in which an OH and Br units sit on adjacent carbon atoms is called a **Halohydrin**.

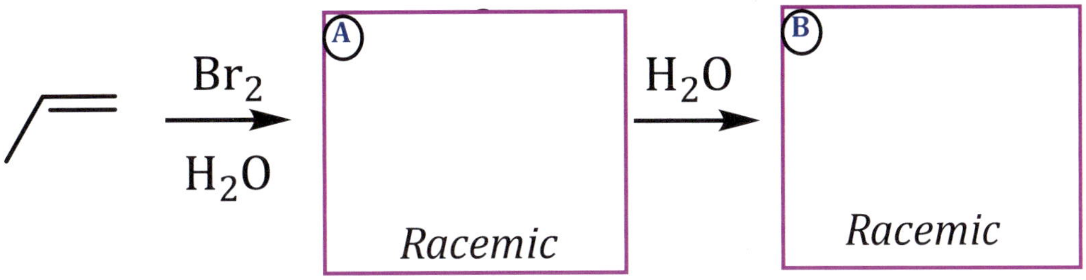

Note that since the water attacks the more substituted side, this is a **Markovnikov product**. The water attacks from the backside of the leaving group Br, so the OH and the Br end up **added *anti-*** to one another:

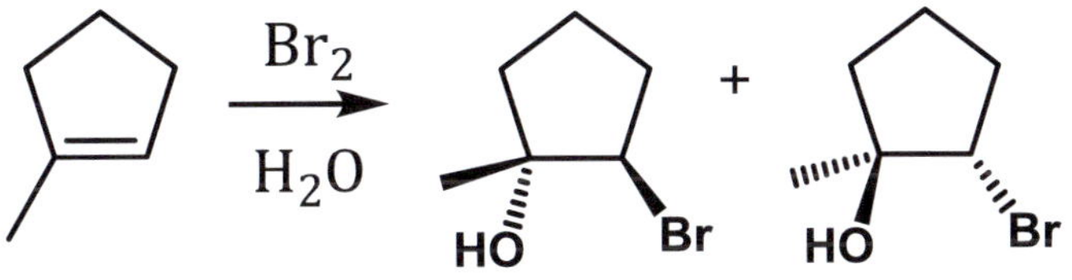

Racemic Mixture

Notes

Lecture Topic III.7: Oxymercuration/Reduction
The Mercurinium Intermediate

Alkenes react with $Hg(OAc)_2$ (mercuric acetate, $^{-}OAc = {}^{-}OC(O)CH_3$) to form a mercurinium intermediate:

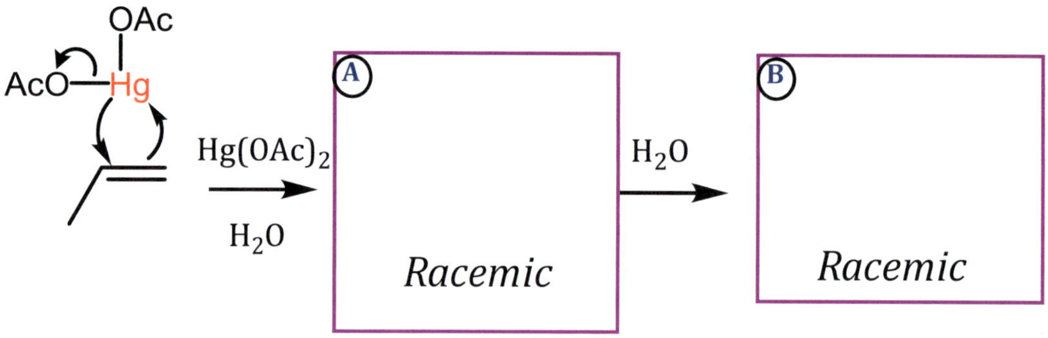

You should recognize that the arrow-pushing to accomplish formation of the mercurinium intermediate in Box A is the same as for forming the halonium ions. Also, in the presence of water, the ring is opened by backside nucleophilic attack at the more substituted carbon in an S_N2 fashion.

Notes

Lecture Topic III.7: Oxymercuration/Reduction

Oxymercuration-Reduction to make Markovnikov Alcohols

The reaction of an alkene with $Hg(OAc)_2$ in the presence of water is called **oxymercuration**. One very useful second step is to replace the mercury with a hydrogen, so that the final species is an alcohol.

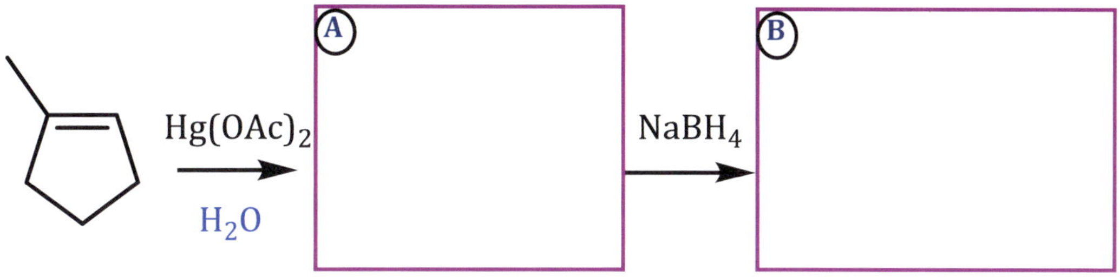

The net result is a **Markovnikov product.** The reduction step has a complex mechanism that is **not specific for *syn*- or *anti*- addition** (the mechanism is shown in the Reaction Guide for this course). The Oxymercuration-reduction sequence is usually written on one arrow as:

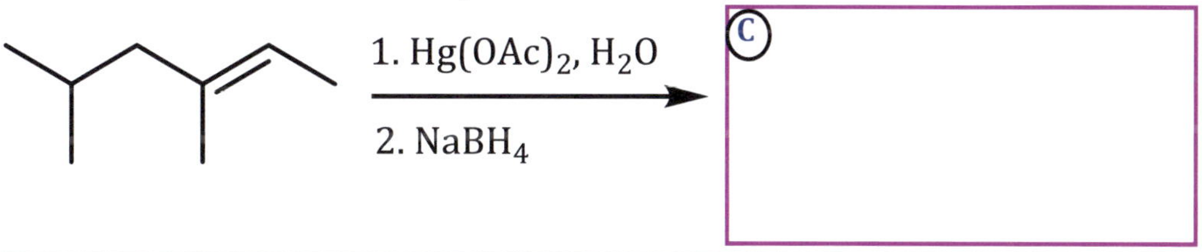

Notes

Lecture Topic III.8: Cyclopropanation and Epoxidation

The Carbene and Cyclopropanation

We have seen how a halogen can add to an alkene to form a cationic three-membered ring. A highly reactive neutral species, called a **carbene** can react with alkenes to make a cyclopropane ring.

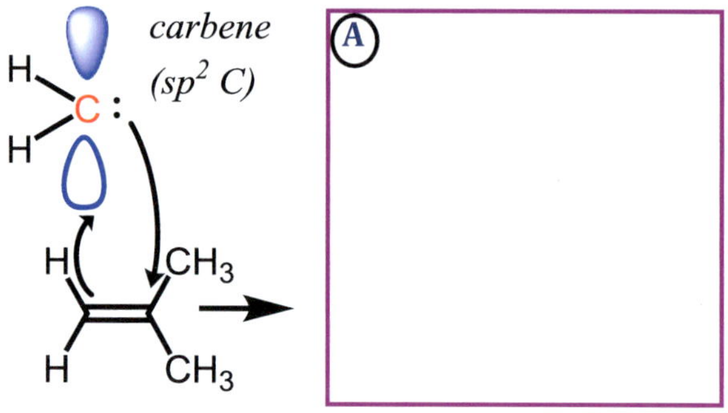

Cyclopropanation is concerted, so the relative distribution of substituents in the alkene is maintained:

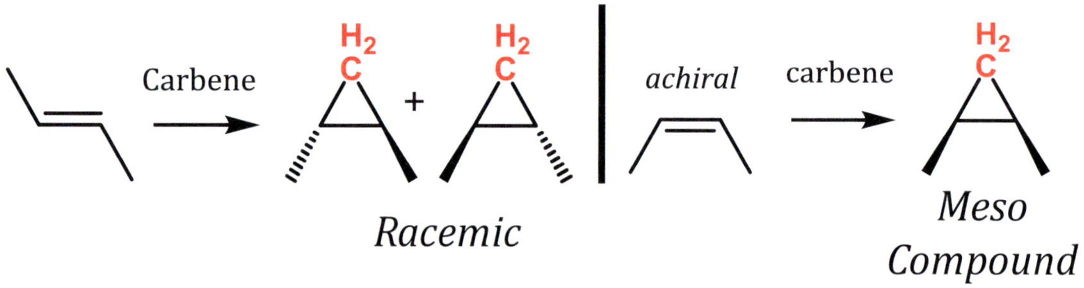

Notes

Lecture Topic III.8: Cyclopropanation and Epoxidation

Three ways to Generate Carbenes for Cyclopropanation

We will learn three ways to generate the carbene or carbenoid (species that reacts as if it were a true carbene) for use in Cyclopropanation:

From Diazomethane

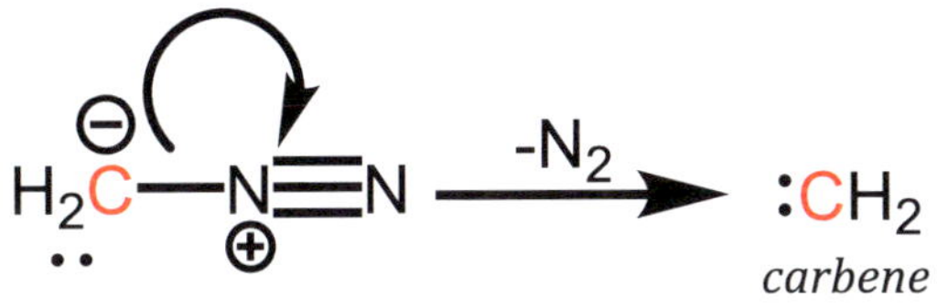

From Chloroform

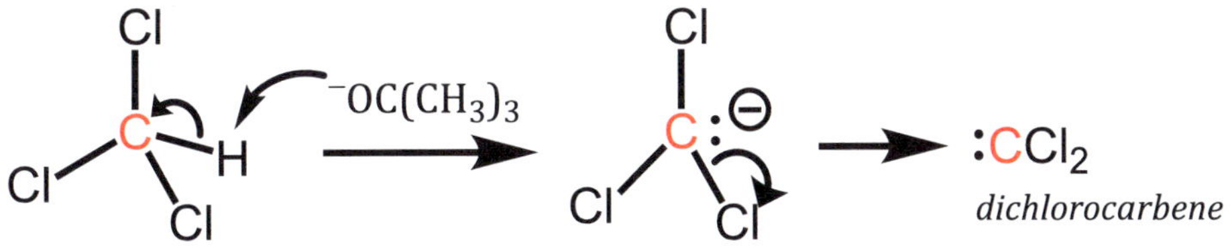

From Diiodomethane: for **Simmons-Smith Cyclopropanation**

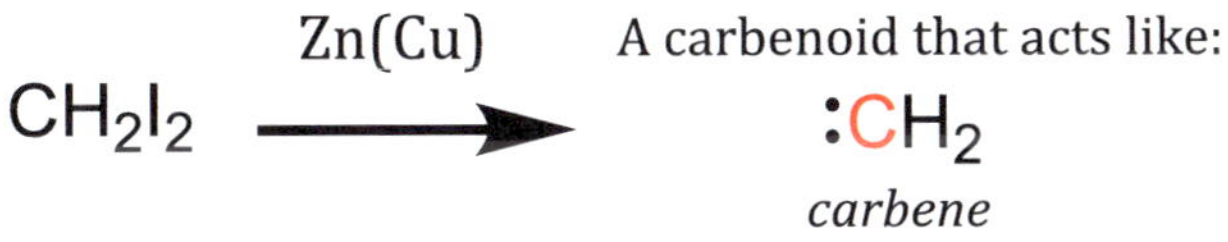

Notes

Lecture Topic III.8: Cyclopropanation and Epoxidation
Cyclopropanation

The three variations of Cyclopropanation are generally written out as shown below. Note the various stereochemical outcomes.

<u>From Diazomethane</u>

A

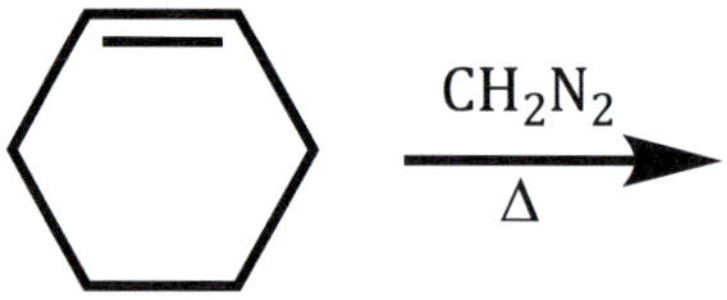

B

<u>From Chloroform</u>

C

<u>Simmons-Smith Cyclopropanation</u>

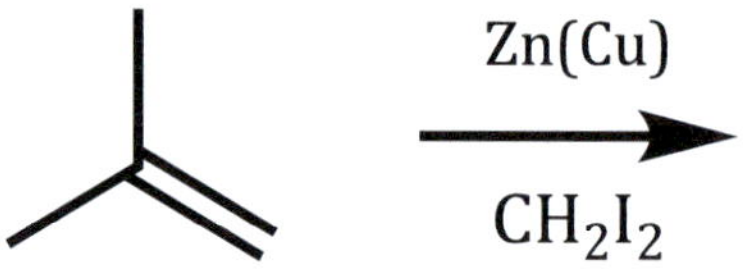

<u>Notes</u>

Lecture Topic III.8: Cyclopropanation and Epoxidation

Epoxidation

Epoxidation is another **concerted** reaction that leads to the formation of a three-membered ring from an alkene. This reaction uses a **peroxyacid** (RC(O)OOR) as the reagent. A common peroxyacid used is mCPBA.

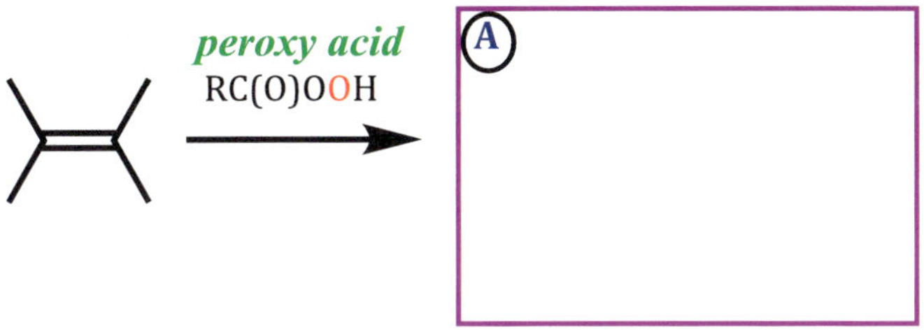

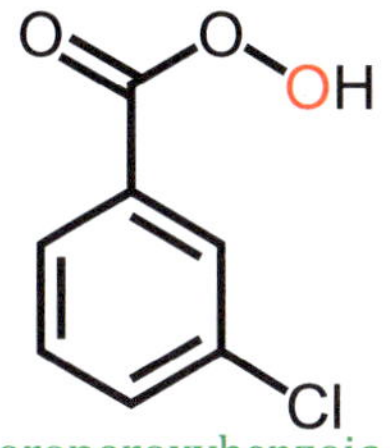

m-chloroperoxybenzoic acid (mCPBA)

The mechanism:

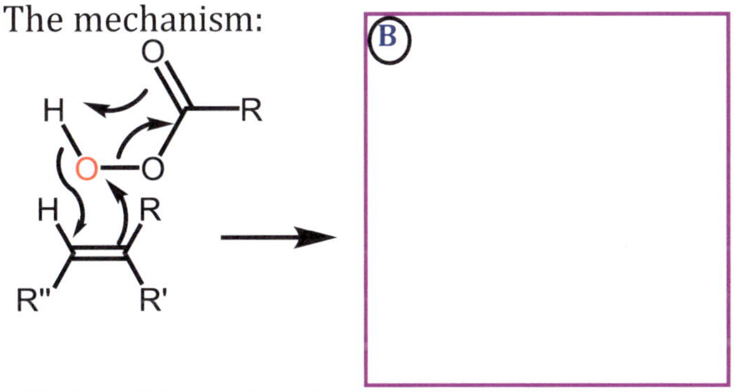

Notes

Lecture Topic III.8: Cyclopropanation and Epoxidation

Epoxidation

Epoxidation of the same three alkenes we looked at for our Cyclopropanation reactions, leads to identical stereochemical outcomes:

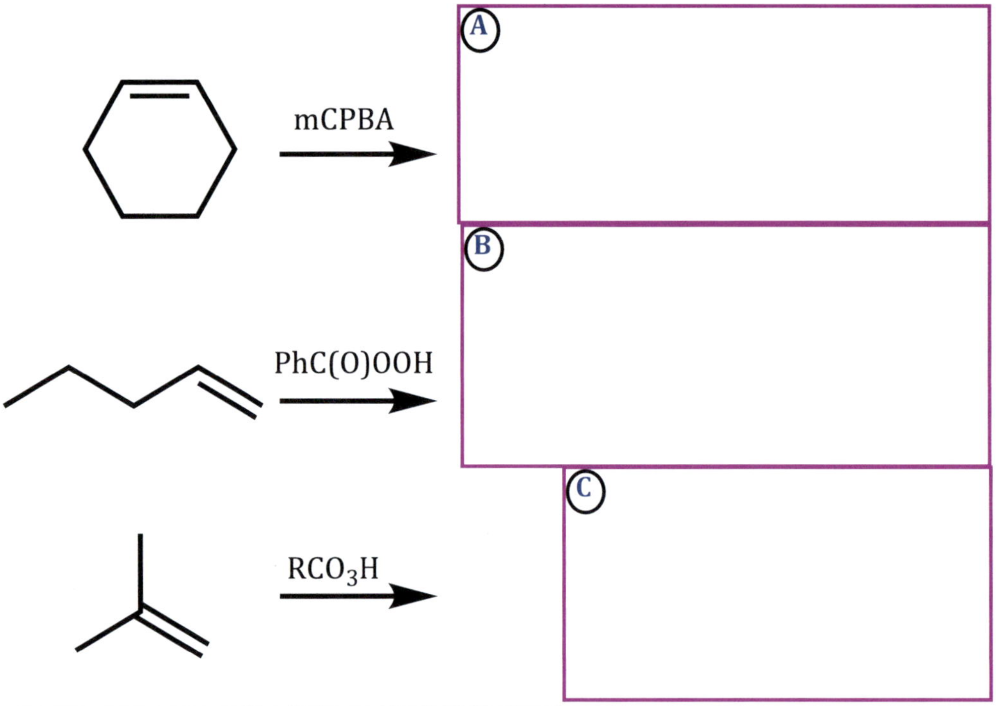

Notes

Lecture Topic III.9: Hydroboration/Oxidation

Borane and Hydroboration

Like a carbene, a **borane** (BH_3, RBH_2 or R_2BH) has an empty *p* orbital, allowing it to accept a pair of electrons from an alkene. Unlike a carbene, borane does not have a lone pair, so it donates a bonding pair from its B–H bond, leading to **hydroboration** of the alkene:

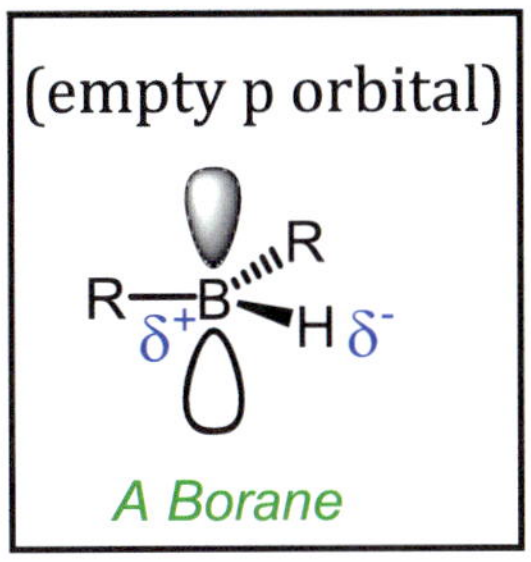

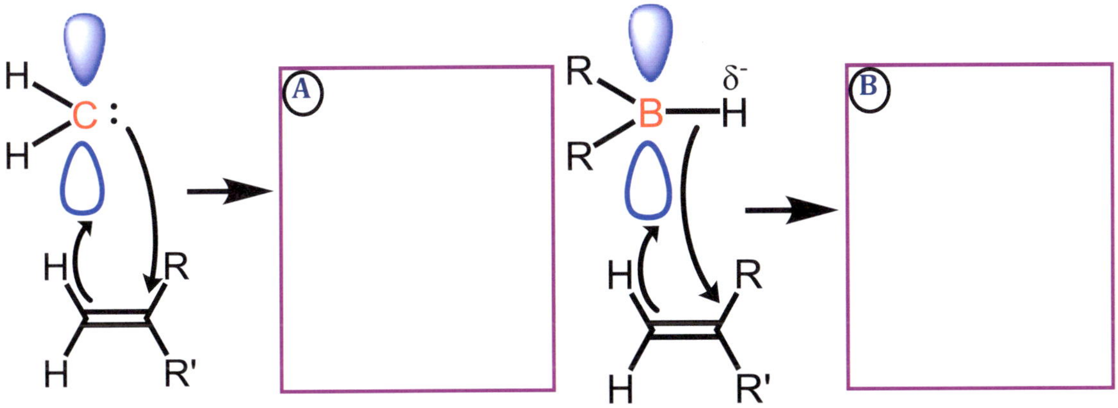

Notes

The **hydroboration step** is concerted and it leads to a **Markovnikov product** (H is more electronegative than B, so it ends up on the more substituted side). Because both B and H add at the same time, this is also a ***syn*-addition** step.

We can understand why H adds to the more substituted side if we examine the transition state and compare it to what we know about halonium formation (**Lecture Topic III.6**)

The more substituted C of the C=C bond better stabilizes positive charge, so the negative charge is attracted to that site

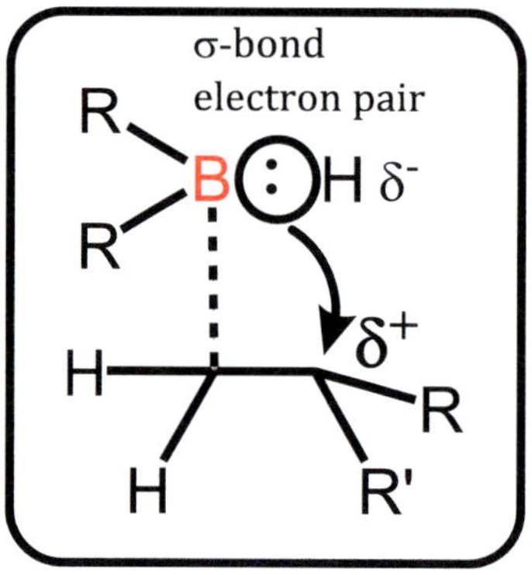

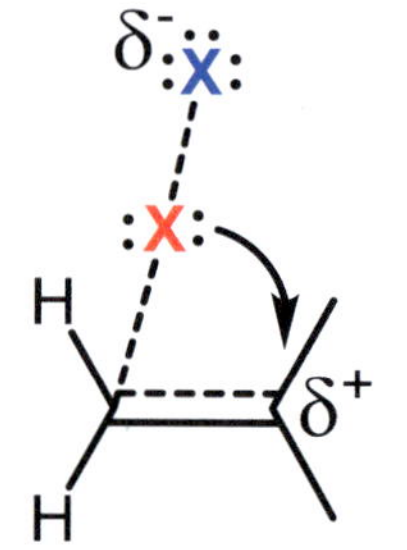

halonium formation transition state

Notes

Lecture Topic III.9: Hydroboration/Oxidation

Hydroboration/Oxidation

The only application of hydroboration in this course is as part of a two-step process called **hydroboration-oxidation.** In this process the hydroboration is followed by an oxidation step in which the borane unit on the carbon is replaced with an OH without changing stereochemistry of attachment:

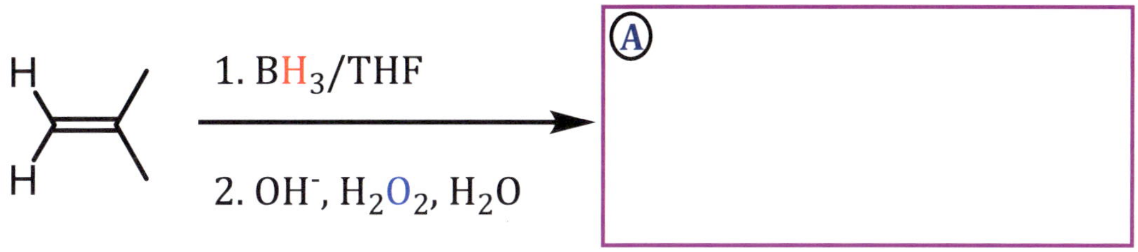

You may see "B_2H_6" or "R_2BH" (such as disiamylborane) in place of "BH_3/THF"; these are alternative hydroboration agents and the net result is the same. The mechanism of the oxidation is complex, but a proposed mechanism is provided in the Reaction Guide.

Notes

Lecture Topic III.9: Hydroboration/Oxidation

Hydroboration/Oxidation to make Non-Markovnikov Alcohols

Because hydroboration places the B on the less substituted C and the B is replaced by an OH in the oxidation, the **net result after both steps** is formation of:

Ⓐ

Because hydroboration is syn- and OH replaces the B without a change of stereochemistry of attachment, the H and OH in the alcohol are:

Ⓑ

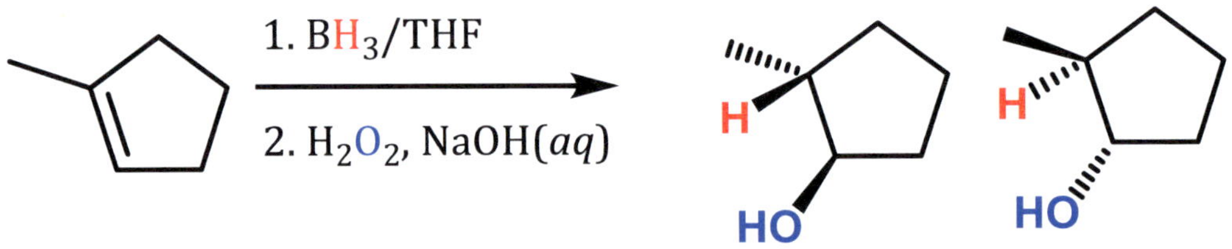

Racemic

Notes

Lecture Topic III.10: Ozonolysis and Preparation of Vicinal Diols

Ozonolysis

Ozonolysis differs from the other reactions of alkenes in that instead of *adding* atoms to the pi bond, ***both* bonds of the C=C are *broken***. The first step of ozonolysis is to react the alkene with ozone (O_3):

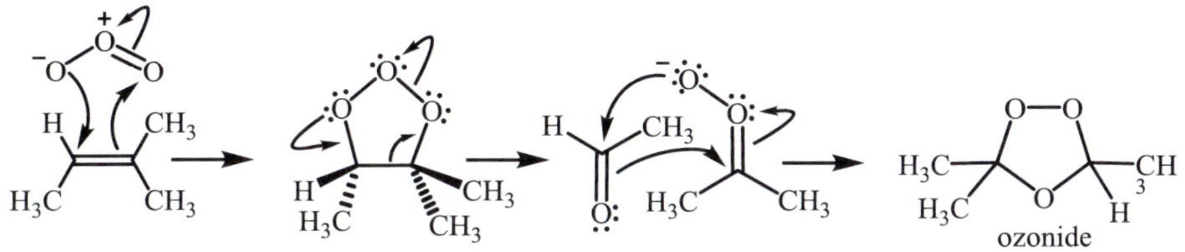

The ozonide is exposed to either reducing or oxidizing workup:

(A) Reducing workup (Zn/H_3O^+ or R_2S):	(B) Oxidizing workup (H_2O_2)

Notes

Lecture Topic III.10: Ozonolysis and Preparation of Vicinal Diols

Ozonolysis

The net ozonolysis reactions that you see generally look like these:

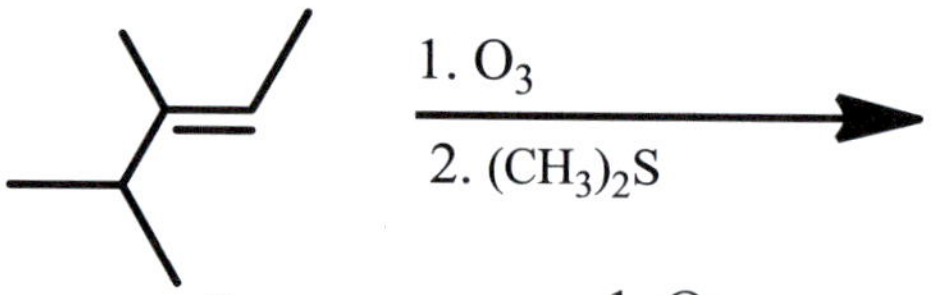

(A)

(B)

If there is more than one double bond, all will react:

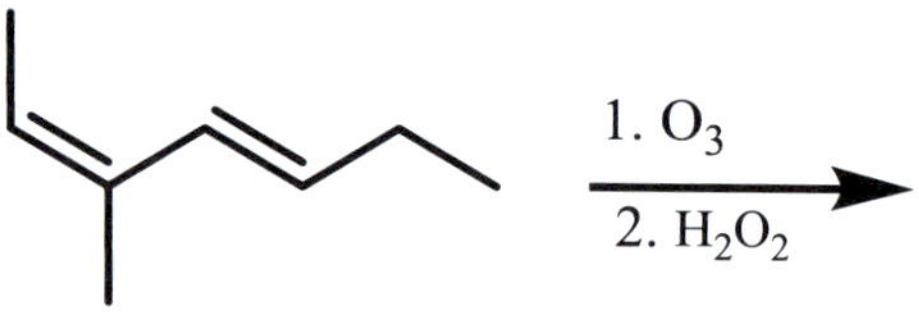

(C)

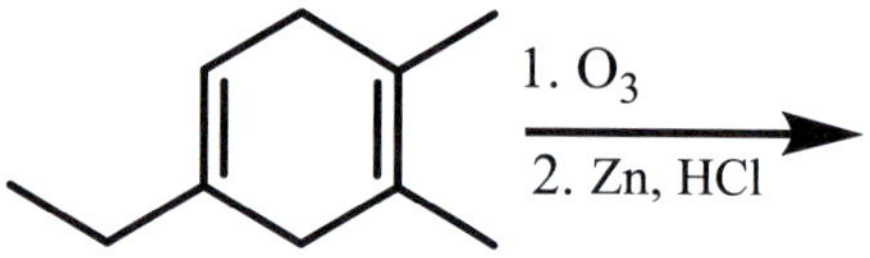

(D)

Notes

Lecture Topic III.10: Ozonolysis and Preparation of Vicinal Diols

Preparation of Vicinal Diols using Osmium Tetroxide

Organic Chemist Definition of oxidation and reduction:

(A)

Reaction of an alkene with osmium tetroxide (OsO_4) followed by a reductive workup leads to net *syn*-addition of two OH groups to the two carbons that were in the C=C bond:

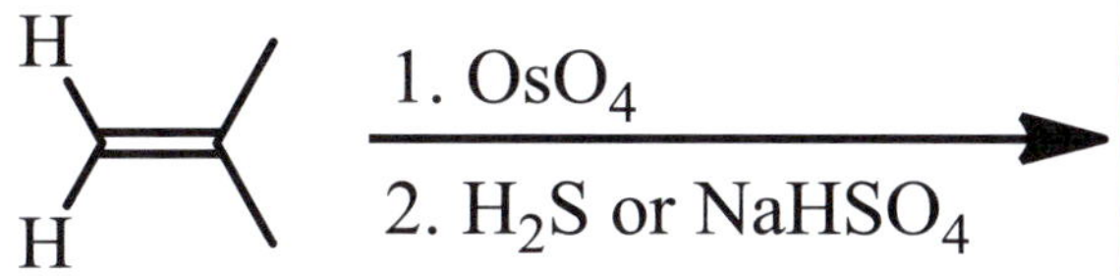

(B)

An example where *syn*-addition would be evident:

(C)

__Notes__

Lecture Topics III.3-10: Recap of Alkene Reactions
Practice, Study and Rationalize

Example. Fill in the major product of each reaction. Show stereochemistry where applicable.

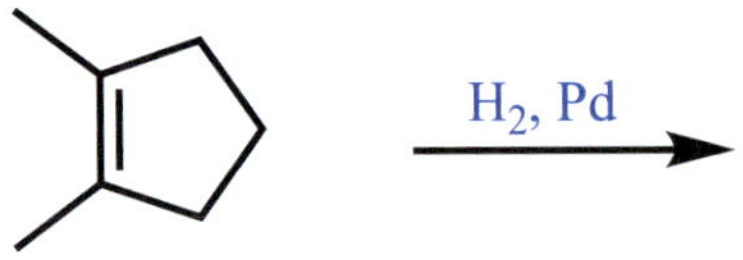

A

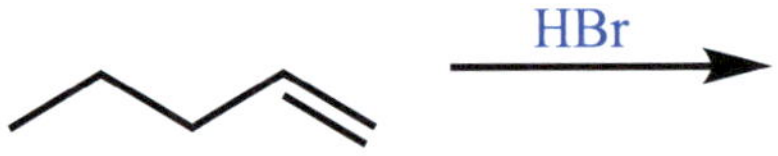

B

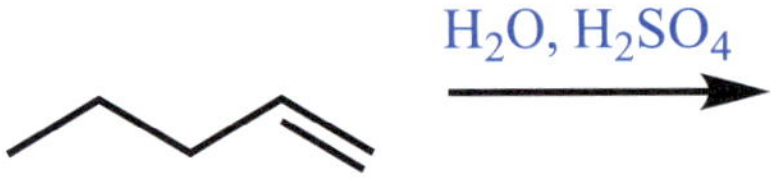

C

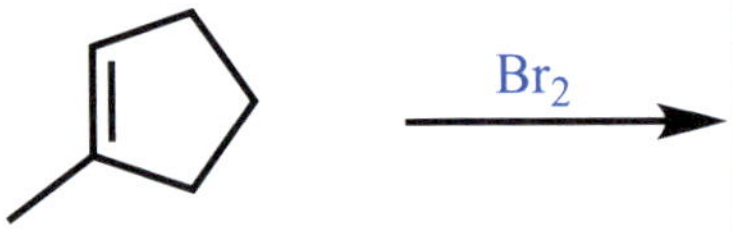

D

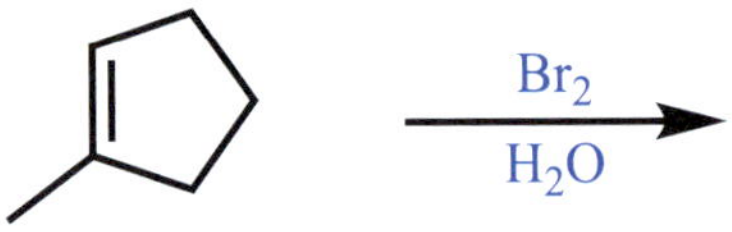

E

Notes

Lecture Topics III.3-10: Recap of Alkene Reactions

Practice, Study and Rationalize

Example. Fill in the major product of each reaction. Show stereochemistry where applicable.

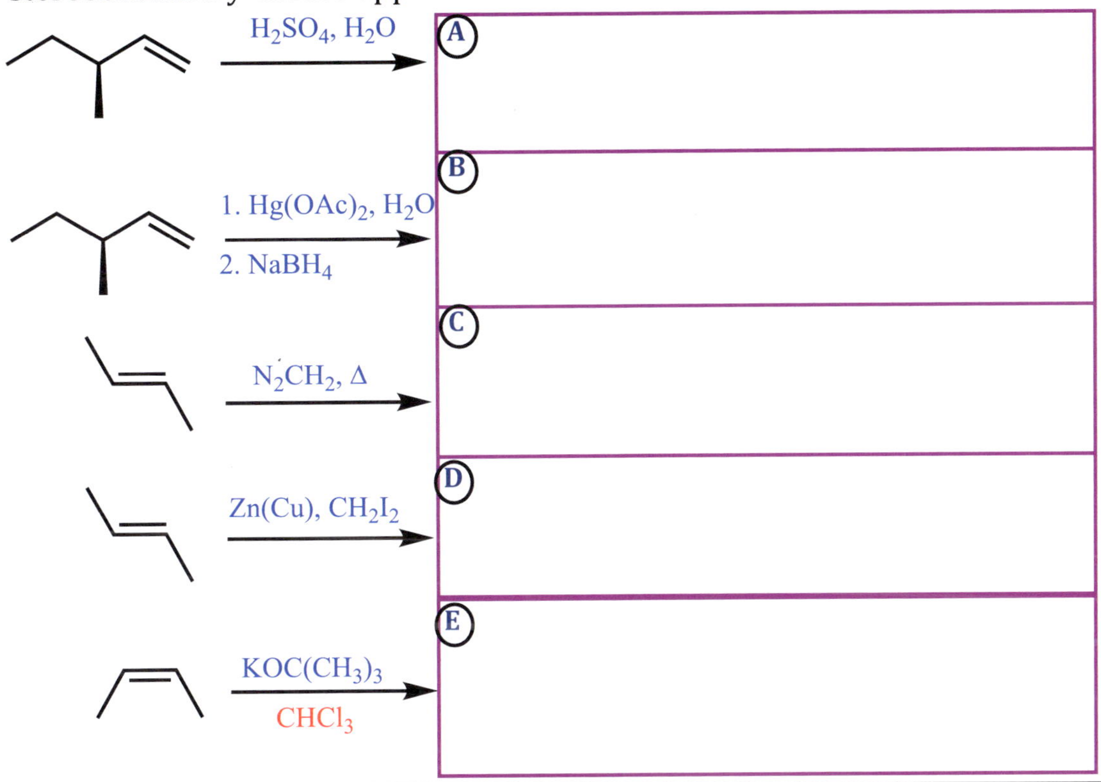

Notes

Lecture Topics III.3-10: Recap of Alkene Reactions
Practice, Study and Rationalize

Example. Fill in the major product of each reaction. Show stereochemistry where applicable.

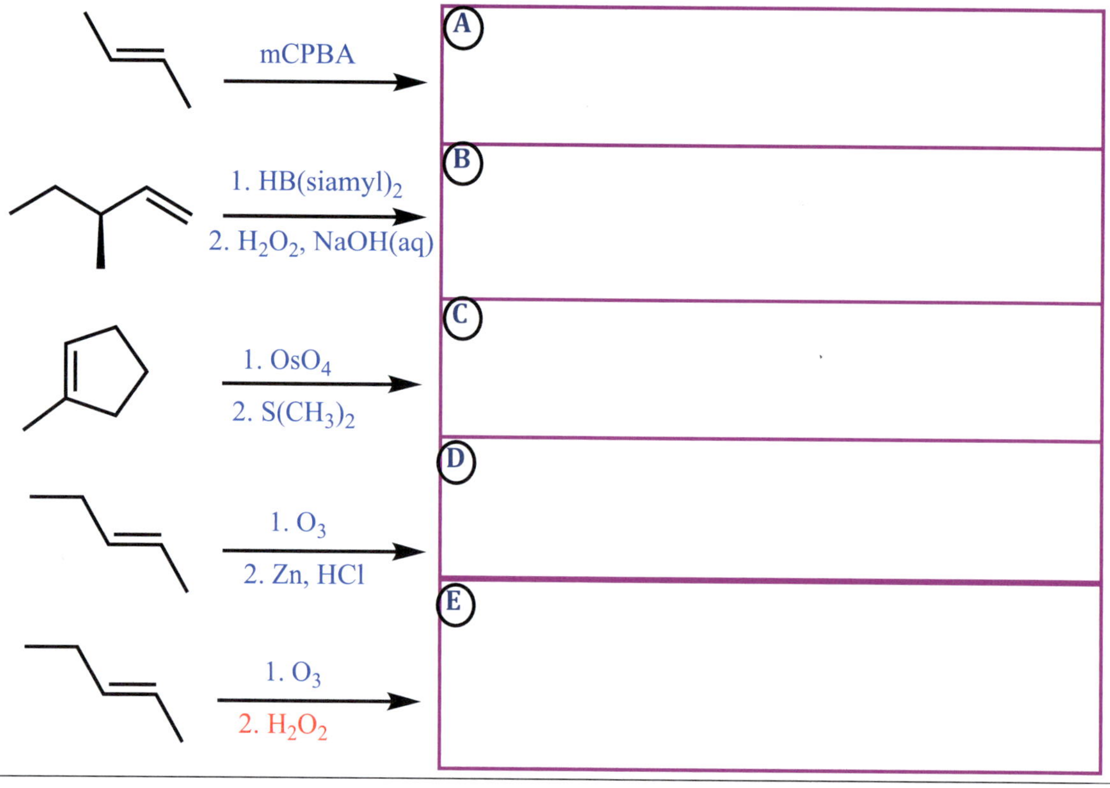

Notes

Lecture Topic III.11: Nomenclature III - Alkynes

Naming Simple Alkynes

acetylene   terminal alkyne   internal alkyne

(A)

Alkanes: **saturated** hydrocarbons

Alkenes: **unsaturated** hydrocarbons (double bond in an alkane)

Alkynes: **unsaturated** hydrocarbons (triple bond in an alkane)

Nomenclature of alkynes is much like that for alkanes, with modifications:

1. always designate the longest chain having

   (B)

2. Number the parent chain such that

   (C)

3. change the 'ane' ending of the name you would use if it were an alkane into an 'yne' ending to indicate that it has a triple bond in it.

__Notes__

Lecture Topic III.11: Nomenclature III - Alkynes

Naming Simple Alkynes

propyne

(A)

(B)

1-pentyne

2-pentyne

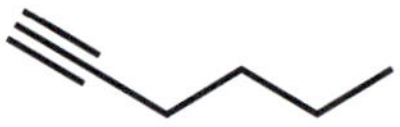
1-hexyne

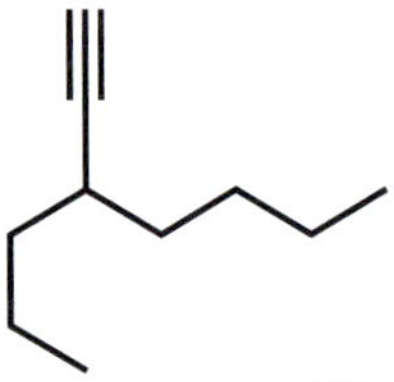
(C)

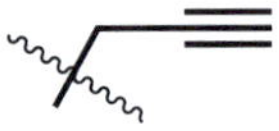
propargyl group

NOTE: Because it is linear there are no E or Z isomers

Notes

Lecture Topic III.11: Nomenclature III - Alkynes

Multiple Alkyne Units, Alkynes with Alcohol or Alkene Units

4. If there is more than one triple bond, use a "-diyne", "-triyne", etc. at the end of the name in place of the "yne", with multiple numbers indicating where the triple bonds are (just like for alkenes):

A

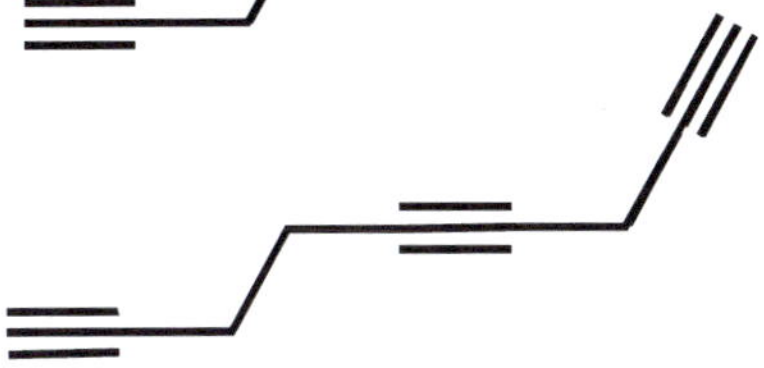

B

5. If alcohol functionalities are present in addition to alkyne:

C

if both alkyne and alkene units are present, number the parent so that

D

IF AND ONLY IF there is a tie between the two in term of numbering the parent, alkene > alkyne (the **lowest** number is given to the one that comes **first** alphabetically).

Notes

Lecture Topic III.11: Nomenclature III - Alkynes
Naming Simple Alkynes

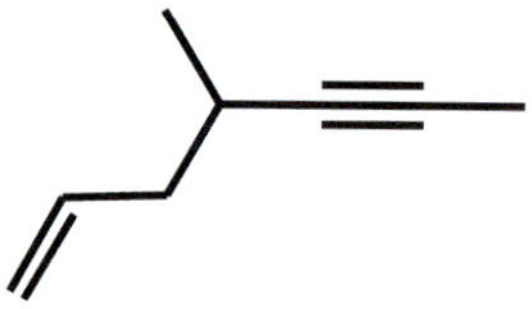

(A)

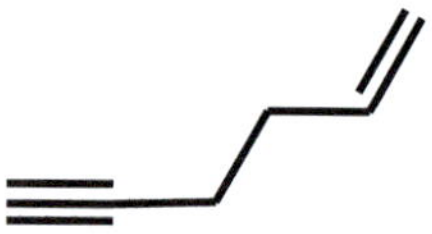

(B)

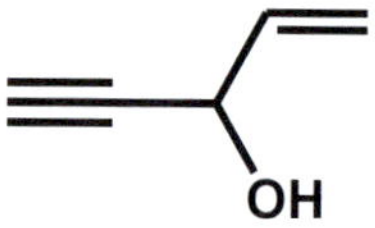

(C)

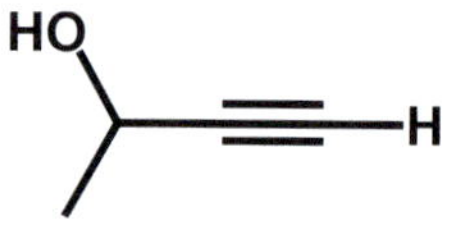

(D)

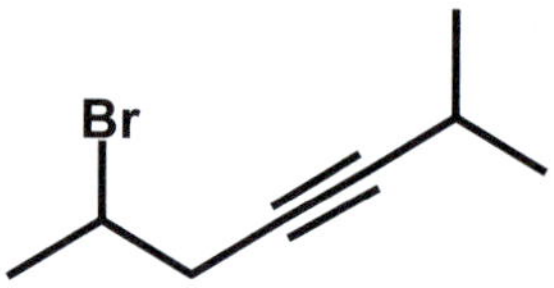

(E)

Note the order of the endings in example C: en-yn-ol

Notes

Lecture Topic III.12: Hydrohalogenation and Halogenation of Alkynes

Preparation of Alkynes by Double Elimination

Alkynes can be prepared by sequential elimination of two equiv HX from an alkyl dihalide via an E2 mechanism, in a manner analogous to that used to prepare alkenes:

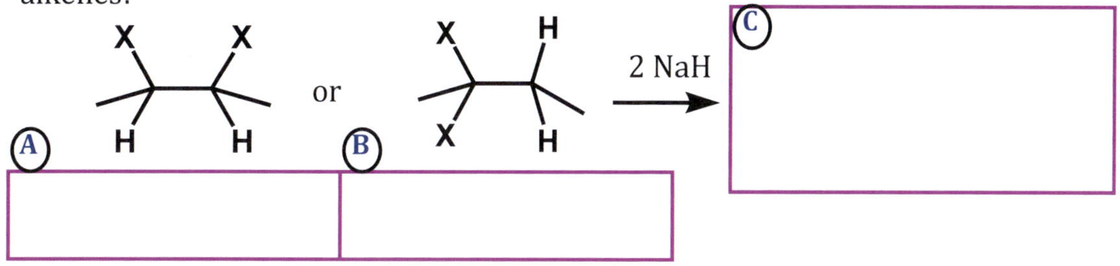

The second elimination is far more difficult, and the reaction can be stopped after one elimination:

The reverse of double elimination is double addition ...

Notes

Lecture Topic III.12: Hydrohalogenation and Halogenation of Alkynes
Addition to Alkyne Pi Bonds

Addition reactions we studied for alkenes can work for alkynes

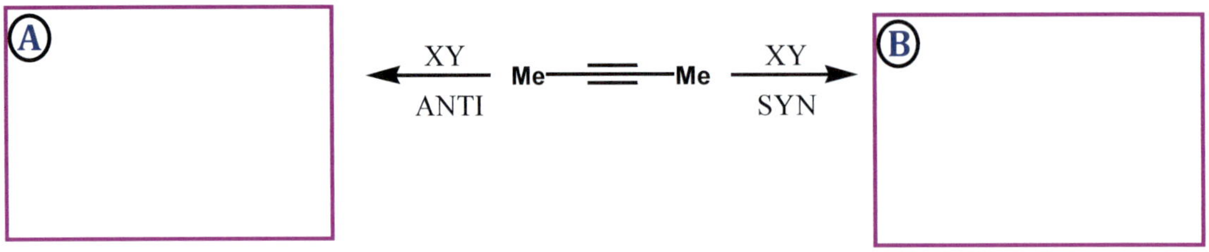

We have TWO pi bonds in an alkyne unit:

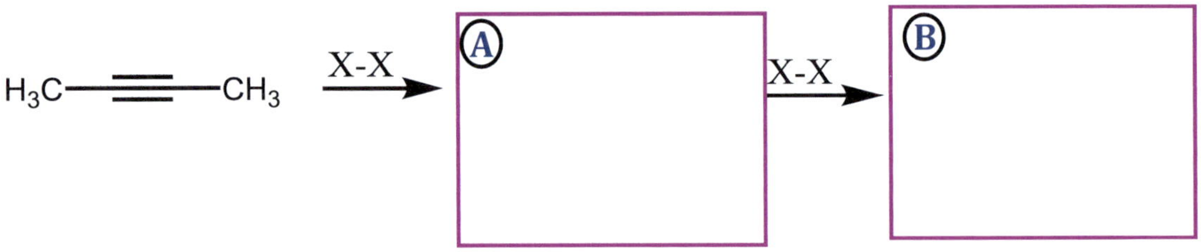

Notes

Lecture Topic III.12: Hydrohalogenation and Halogenation of Alkynes

Hydrohalogenation

Alkynes can react with HX, yielding Markovnikov products. You should be able to deduce the mechanism in analogy to the reaction you know for alkenes. One equiv of HX leads to an alkene. Excess HX leads to the alkane.

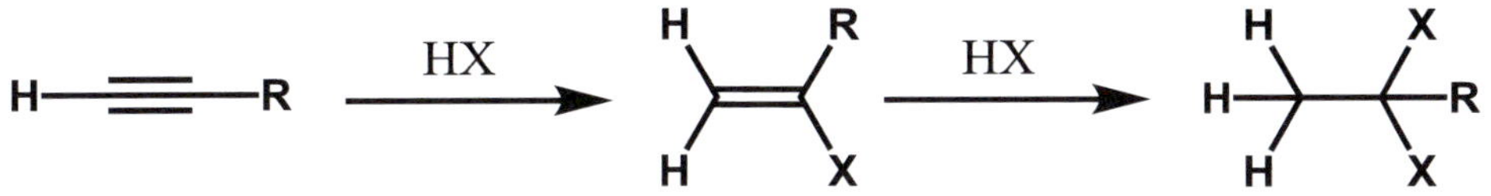

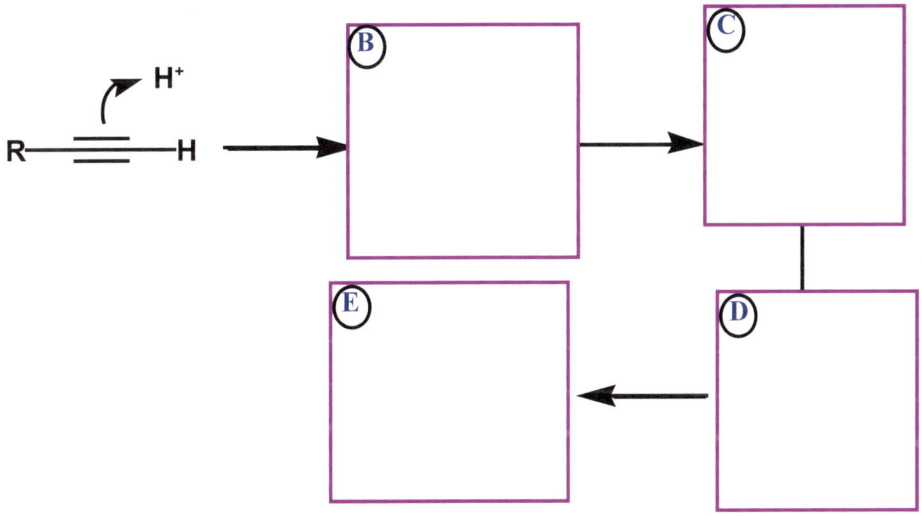

Notes

Lecture Topic III.12: Hydrohalogenation and Halogenation of Alkynes

Halogenation

Alkynes react with halogens in analogy to alkenes. One or two halogen additions can take place across the unsaturated unit:

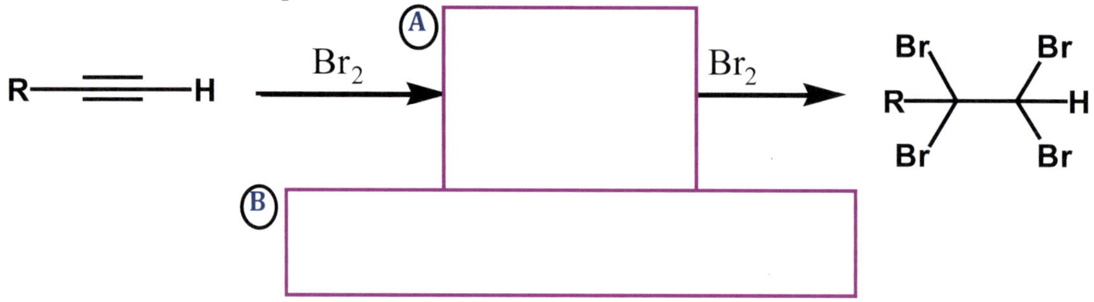

Though the net result is what we saw for halogenation of alkenes, the mechanism is different:

C

__Notes__

Lecture Topic III.13: Three Ways to Reduce Alkynes
Hydrogenation

An alkyne can be hydrogenated using conditions similar to those used to hydrogenate an alkene (H_2/ transition metal catalyst):

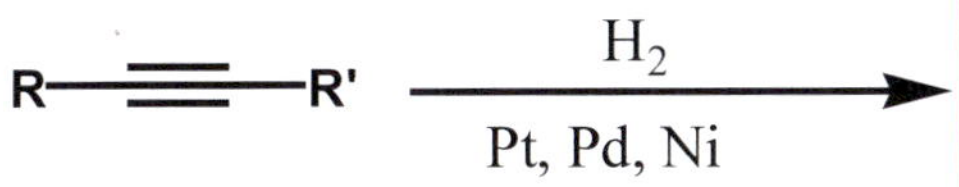

(A)

Gentler conditions (a less active catalyst) are required if conversion of an Alkyne to an alkene is desired without further hydrogenation. To do this, use H_2 with **Lindlar's catalyst** ($Pd/CaCO_3$, $Pb(OAc)_2$, quinoline).

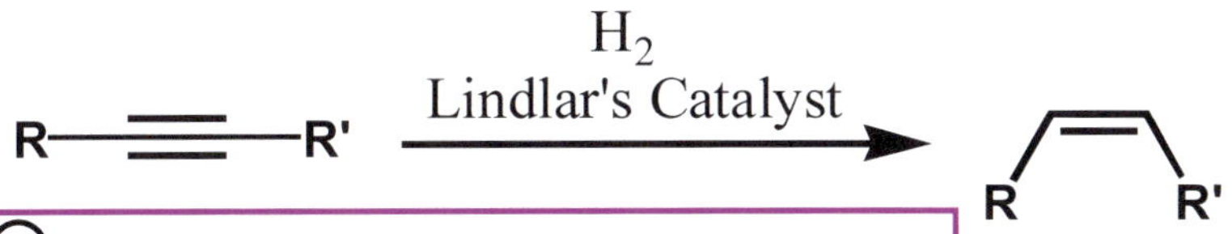

(B)

Notes

Lecture Topic III.13: Three Ways to Reduce Alkynes
Reduction with Na (or K) with NH_3

When an alkali metal such as Na or K is dissolved in liquid ammonia, solvated electrons are formed. The resulting solution can be used to reduce alkynes:

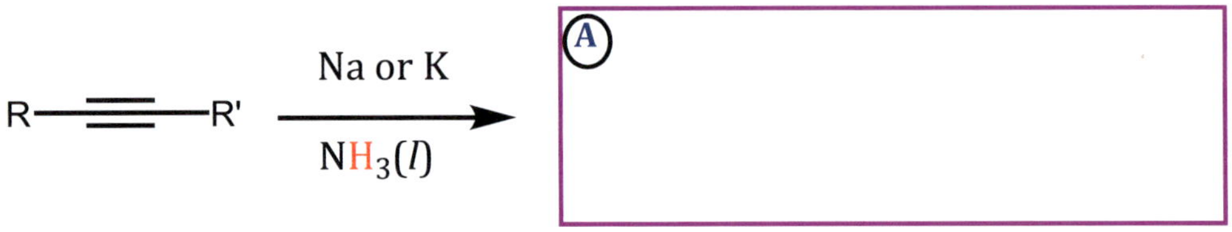

The mechanism is somewhat complex and involves radicals:

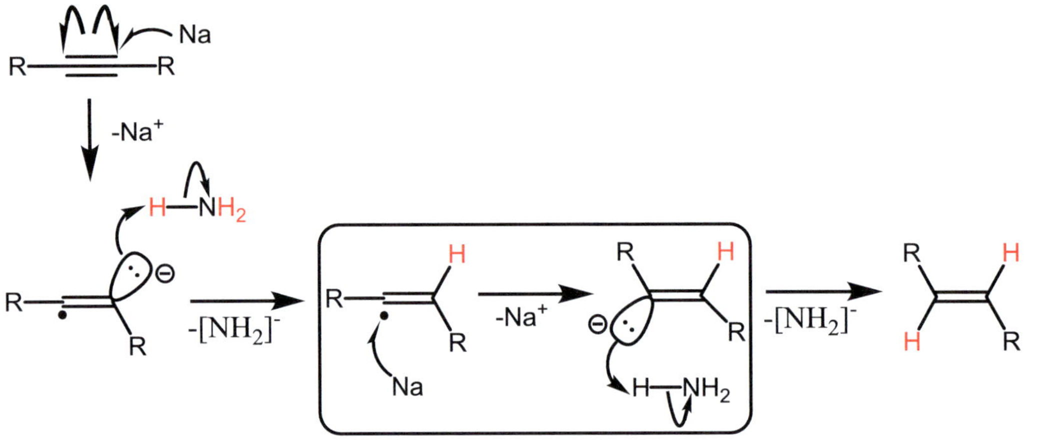

Notes

Lecture Topic III.14: Facile Constitutional Isomerism
Tautomerization

Tautomers:

(A)

(B)

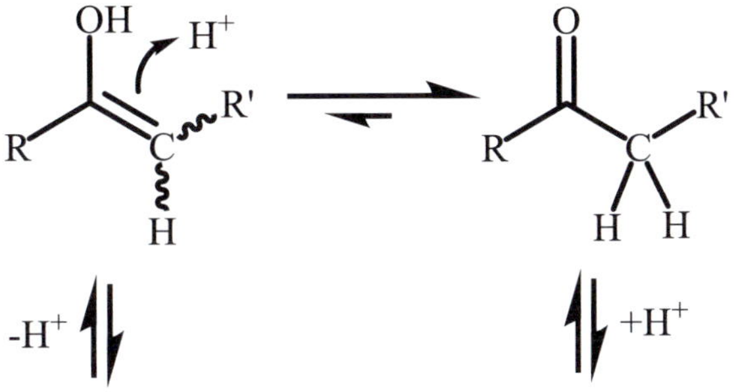

(C)

(D)

Notes

Lecture Topic III.15: Preparation of Carbonyls from Alkynes

Hydration in the Presence of Mercury Salt

Alkynes can undergo hydration reactions:

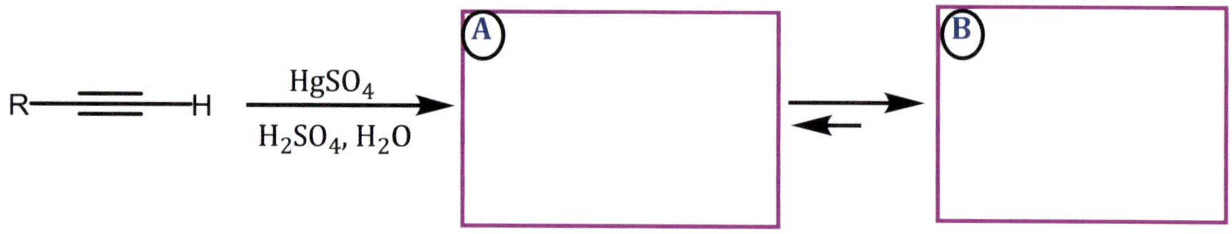

As with alkenes, hydration of an alkyne leads to initial formation of a Markovnikov alcohol. This tautomerizes, so the observed product will be a carbonyl.

Another difference fromalkene hydration is that e need to add a mercury salt. This is needed because:

C

Notes

Lecture Topic III.15: Preparation of Carbonyls from Alkynes
Hydroboration/Oxidation

Alkynes can undergo hydroboration/oxidation:

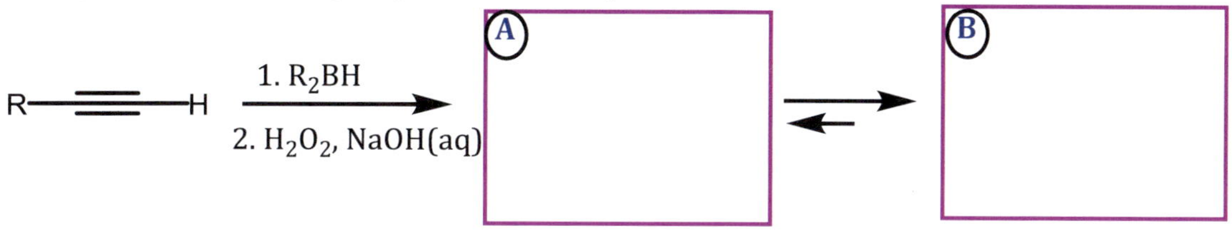

As with alkenes, hydroboration/oxidation of an alkyne leads to initial formation of a Non-Markovnikov alcohol. This tautomerizes, so the observed product will be a carbonyl.

The hydroboration/oxidation of a terminal alkyne leads to formation of an aldehyde:

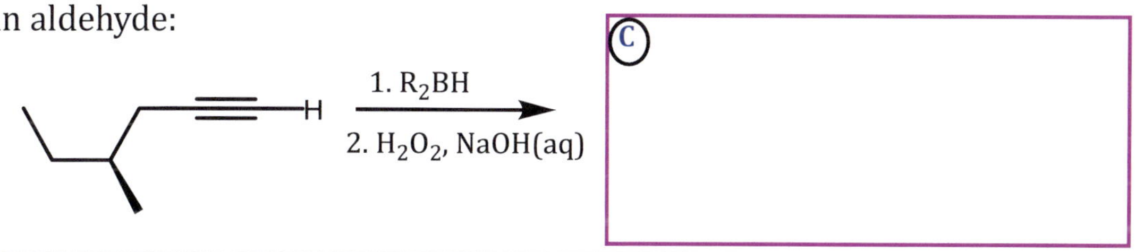

__Notes__

Lecture Topic III.16: Acetylide Anion Preparation and use as a Nucleophile

Chain Extension of Terminal Alkynes

A **strong** base can deprotonate a terminal alkyne:

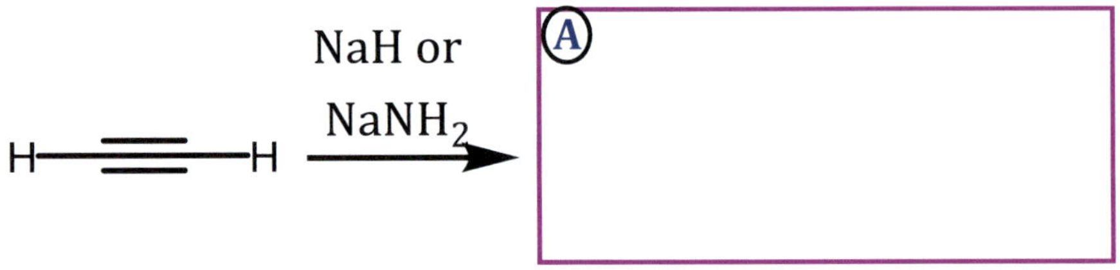

The **acetylide anion** can then be used as a nucleophile in an S_N2 reaction:

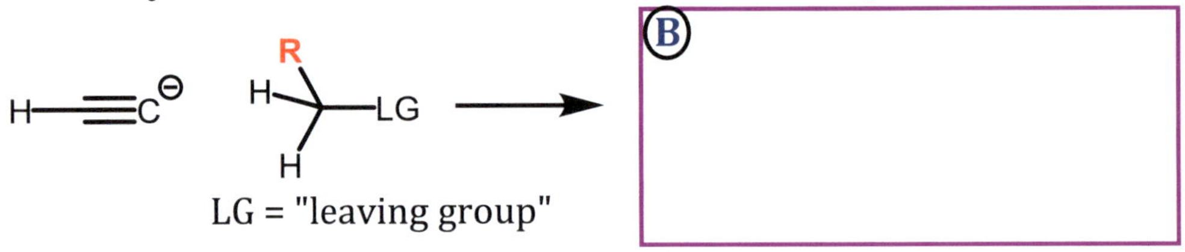

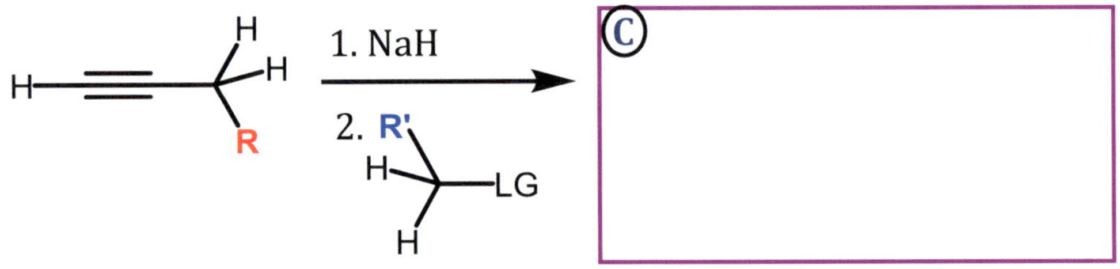

Notes

Lecture Topic III.17: Radicals I

Peroxide-Mediated Hydrobromination

Peroxide-initiated hydrobromination is an alternative way to add H and Br

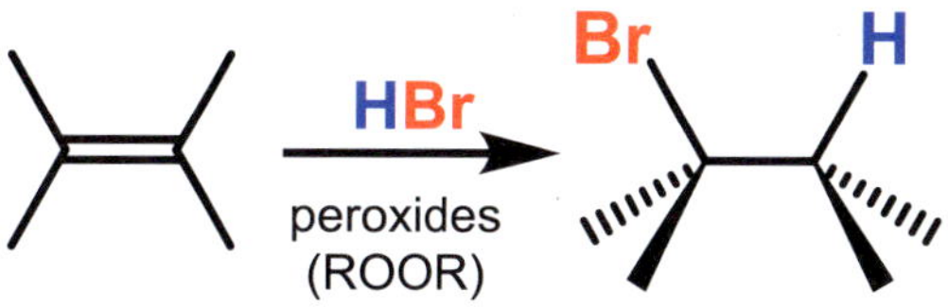

Ⓐ **Initiation**

Ⓑ **Propagation**

Ⓒ **Net Reaction**

Notes

Lecture Topic III.17: Radicals I
A Radical Chain Reaction

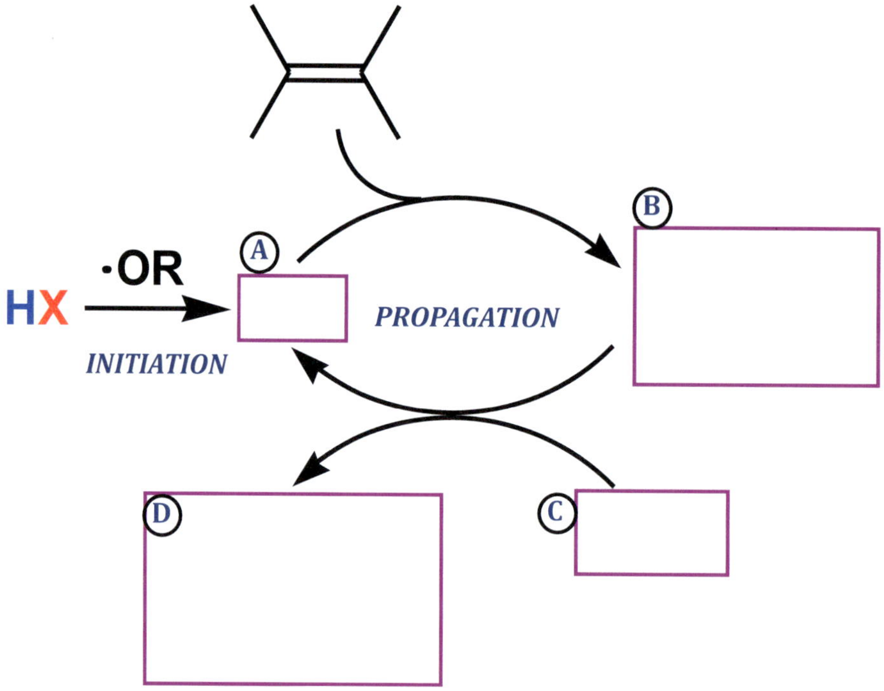

Termination occurs when any two radicals come together:

Notes

Lecture Topic III.17: Radicals I

Peroxide-Mediated Hydrobromination

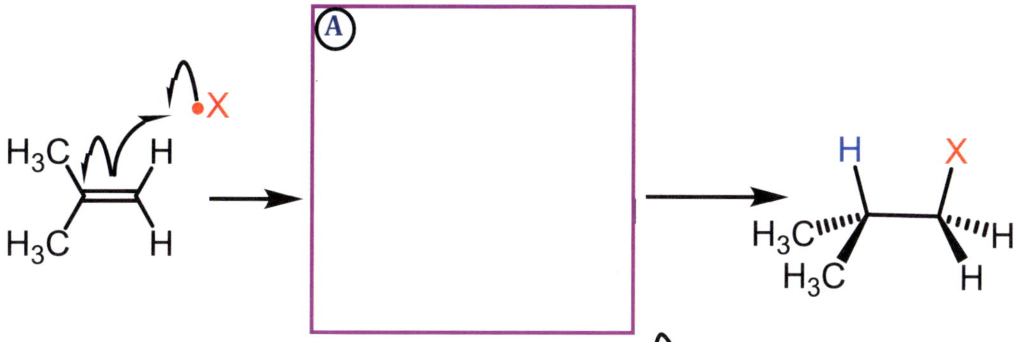

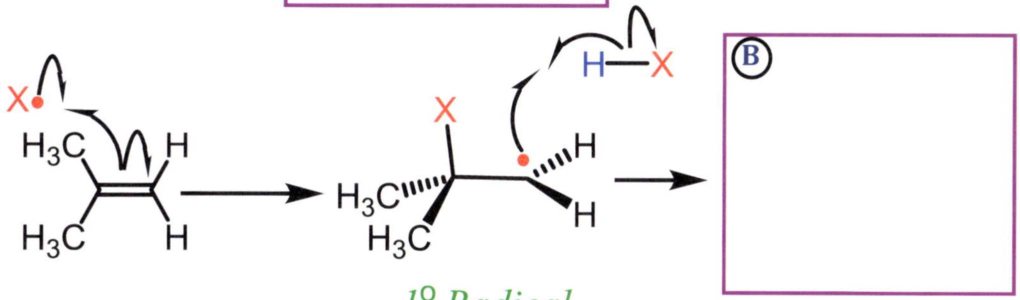

1° Radical

Once again, the major product is that derived from the more stable intermediate, in this case the more substituted radical that picks up the H atom, so we get: (C)

Notes

Lecture Topic III.17: Radicals I

Peroxide-Mediated Hydrobromination

Note that the peroxide-mediated (radical intermediate) hydrohalogenation yields the so-called "anti-Markovnikov" product, while the peroxide-free hydrohalogenation (carbocation intermediate) yields the Markivnikov product. This gives a researcher access to whichever product is desired:

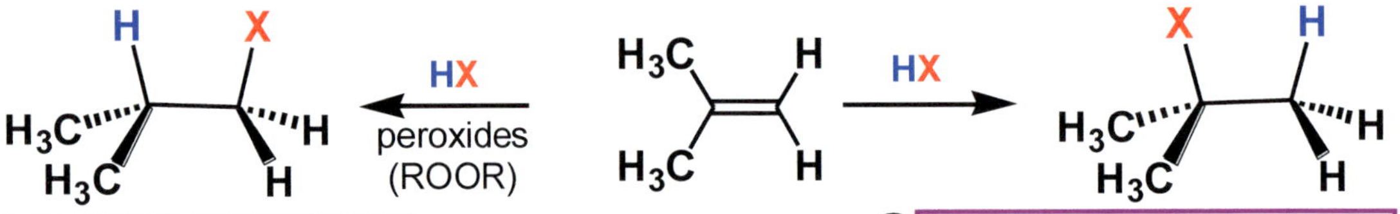

A

B

C

Notes

Lecture Topic III.18: Radicals II
Radical Halogenation of Alkanes

Alkyl halides can also be prepared from alkanes. Alkanes do not have very polar bonds, and do not readily react with common acids or bases. If we generate a very reactive species, like a **radical** we can observe reactions of alkanes. A good way to generate a halide radical is by excitation of X_2 with light.

$$R\text{–}H + X_2 \xrightarrow[\text{(light)}]{h\nu} R\text{–}X + HX$$

The dihalides have different reactivities towards radical formation, so only **Cl_2 and Br_2** are commonly used in the above reaction:

$$I_2 < \boxed{Br_2 < Cl_2 <} F_2$$

REACTIVITY

Because we are using such reactive species (radicals) it is often difficult to control these reaction to yield a single product, as we will see ...

Notes

Lecture Topic III.18: Radicals II
Radical Halogenation of Alkanes

We can write the discrete steps of the **radical chain mechanism**:

Ⓐ **Initiation**

Ⓑ **Propagation**

Ⓒ **Net Reaction**

Ⓓ **Termination**

Notes

Lecture Topic III.18: Radicals II
Radical Halogenation of Alkanes

... or we can write a simple cycle (shows **initiation** and **propagation** but not **termination**):

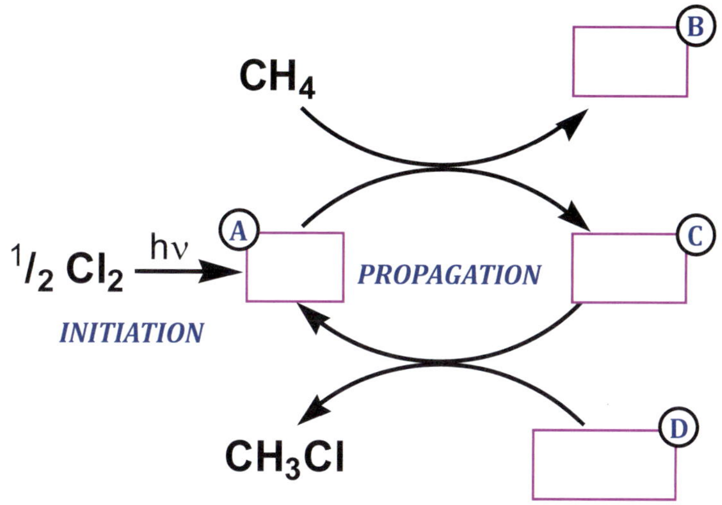

Termination occurs when any two radicals come together.

Notes

Lecture Topic III.18: Radicals II
Radical Halogenation of Alkanes

Only **Cl_2 and Br_2** are commonly used in this radical chain reaction because fluorine is too reactive and iodine is not reactive enough. The structure of the alkane we use also plays an important role.

The key step is the homolytic cleavage of the alkane C-H bond to form the radical:

$$R–H \rightarrow R\bullet + H\bullet \quad \Delta H = \text{BDE}$$ (**bond dissociation enthalpy**)

Easier radical formation ↓

BDE Trends:

Structure	BDE	
CH_3—H	104 kcal	
CH_3CH_2—H	98 kcal	
$CH_3CH_2CH_2$—H	98 kcal	(1º)
$(CH_3)_2CH$—H	95 kcal	(2º)
$(CH_3)_3C$—H	91 kcal	(3º)

The take-home lesson is:

(A)

Notes

Lecture Topic III.18: Radicals II
Radical Halogenation of Alkanes

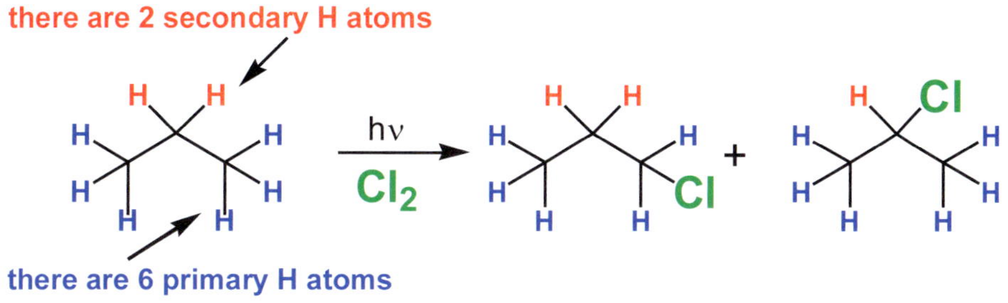

Experimentally observed:	43%	57%
Statistically anticipated:	75%	25%

We observe a greater than statistically anticipated reactivity at the secondary site. The reaction has a greater **regioselectivity** for this site.

We already know that the secondary radical is more stable than the primary, and that the secondary C-H bond has a lower BDE, so this regioselectivity makes sense.

Notes

Lecture Topic III.18: Radicals II
Radical Halogenation of Alkanes

When Cl_2 is used, the relative reactivity can be approximated as:

Relative Reactivity of C–H:	3°	2°	1°
	5.2	3.9	1

Use this to predict product distribution for the chlorination of propane:

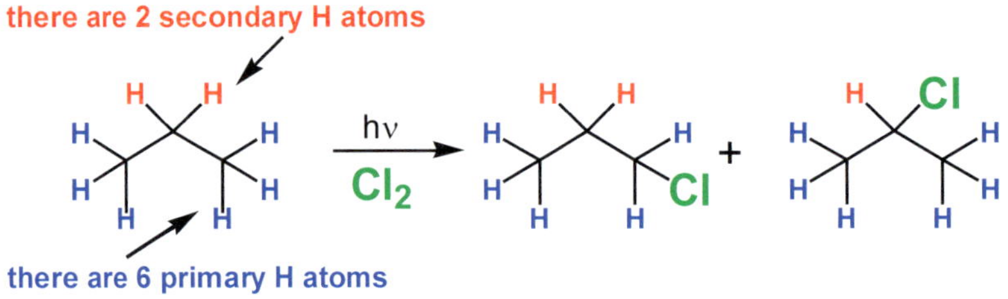

Rel. amount of 1° product = reactivity of 1° H x number of 1° H's
Rel. amount of 2° product = reactivity of 2° H x number of 2° H's

= 3.9 x 2 = 7.8 = 43% primary
= 1 x 6 = 6 = 57% secondary

Notes

Lecture Topic III.18: Radicals II
Radical Halogenation of Alkanes

Bromine is less reactive and therefore more **selective** than chlorine. Consider the observed products for these reactions:

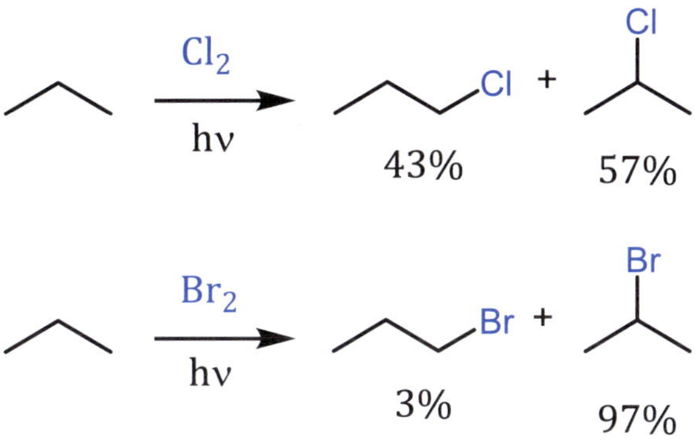

From these data we can deduce that the selectivity of bromine must be more pronounced than for chlorine. This is the case:

Site:	3°	2°	1°
Relative reactivity:	1640	82	1

Notes

Lecture Topic III.19: Introduction to Polymers

Definitions

Alkenes can be used as starting materials to make **polymers**.

POLYMER:

MONOMER:

REPEAT UNIT:

EXAMPLE:

Notes

Lecture Topic III.20: Radicals III

Radical-Initiated Polymerization of Alkenes

Radical Polymerization of an alkene:

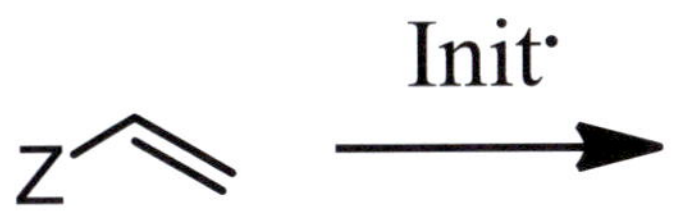

A

Mechanism of Polymer chain growth:

B

Unlike the other radical reactions of alkenes that we have seen, there is ONLY an initiator added to the alkene, no other reagents. A common radical initiator is Benzoyl Peroxide, $(PhC(O)O)_2$

Notes

Cationic Polymerization of an alkene:

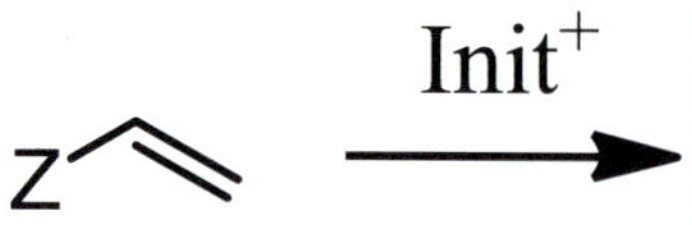

A

B Chain Growth:

Lewis acids can be good initiators for a cationic polymerization:

C

Notes

Anionic Polymerization of an alkene:

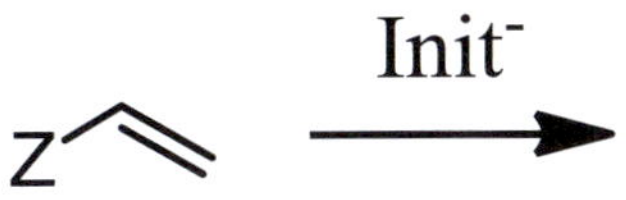

A

B Chain Growth:

You may notice that in the other reactions of alkenes, a cation or radical was the key intermediate. This is because it is more difficult to generate a carbanion. In order to do a successful anionic polymerization, **the group Z must**:

C

Notes

Lecture Topic III.21: Anionic and Cationic Polymerization of Alkenes

Examples of Important Commercial Polymers

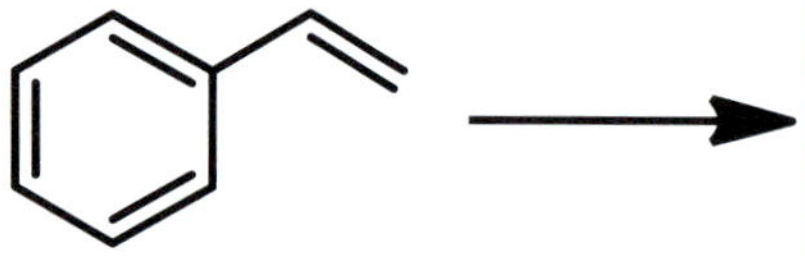

(A) polystyrene

(B)

polyethylene

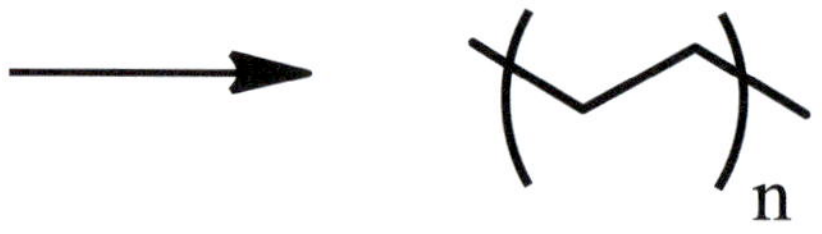

(C)

poly(methyl methacrylate)

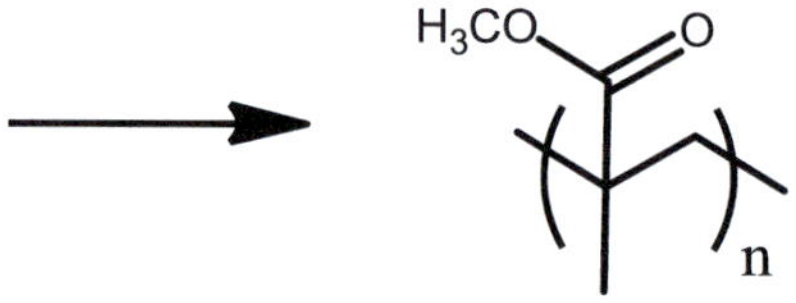

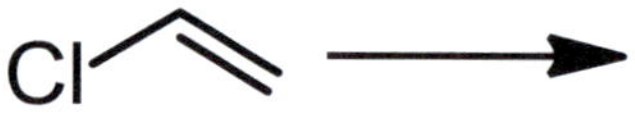

(D) Poly(vinyl chloride) (PVC)

Notes

Index

Made in the USA
Las Vegas, NV
04 November 2026

57424147R10162